It is incredible that so little is understood about the framework of our bodies. When it fails, and the miseries of arthritis and backache occur, evolution is blamed. Evolution has not failed. We are perfectly designed for two-legged posture.

You cannot keep up with the literature produced on hearts and circulation, but are hard pressed to find any authoritative writing on the body's frame-work, or posture. Following forty years' interest, and ten years' research the author has produced a simple, readable explanation of the ramifications of posture on our health and enjoyment of life. Taking the reader through the fascinating evolution of man into the upright posture, he shows that our chairborne society is not only ruining our circulation, but is also having disastrous effects on a variety of other conditions.

He provides the answers to two of our greatest modern curses both in pain and financial cost — osteoarthritis and backache. He shows that the true, mutual sexual satisfaction that nature has designed in sex is impossible with bad posture. New concepts are detailed in maternity, obesity, appearance, personality, eyesight, old age, childhood, industry and sport — all based on posture.

In a book teeming with new and logical ideas the author has added appendices outlining his ideas on how a cell divides (vital knowledge in cancer research): how the brain is electrically recharged: and man as an electric machine.

His enthusiastic wonder of evolution and its greatest product, the human body, shines through in his optimistic explanations of the prevention and treatment of the ill effects of incorrect posture.

POSTURE MAKES PERFECT

POSTURE MAKES PERFECT

by **Dr.Vic Barker**

General Practitioner

JAPAN PUBLICATIONS, INC.

Originally published in Australia and New Zealand
© 1985 Fitworld Publications

Published by JAPAN PUBLICATIONS, INC., Tokyo and New York

Distributors:
UNITED STATES: *Kodansha America, Inc. through Farrar, Straus & Giroux, 19 Union Square
West, New York, N.Y. 10003,* CANADA: *Fitzhenry & Whiteside Ltd., 91 Granton Drive,
Richmond Hill, Ontario, L4B 2N5.* BRITISH ISLES AND EUROPEAN CONTINENT: *Premier Book
Marketing Ltd., 1 Gower Street, London WC1E 6HA.* AUSTRALIA AND NEW ZEALAND:
Bookwise International, 54 Crittenden Road, Findon, South Australia 5023. THE FAR EAST
AND JAPAN: *Japan Publications Trading Co., Ltd., 1–2–1, Sarugaku-cho, Chiyoda-ku,
Tokyo 101.*

First edition: May 1993

LCCC No. 91-076147
ISBN 0-87040-871-2

Printed in Hong Kong by Colorcraft Ltd.
6A, Victoria Centre, 15 Watson's Road, North Point

CONTENTS

FOREWORD

In this decade's outpouring of books promoting a better physical life, it is a relief to read one that has the power and originality to demand a lot more than a brief skim over the surface. Authors in this field usually draw so heavily from their contemporaries that it is hard to find essential differences and if there are any they often exist in the form of a preoccupation with some minor aspect of the physical that can be sensationalised for profit.

Posture Makes Perfect is worthy of serious and in depth study for the author has set forth a series of well reasoned and in some cases astounding propositions for both the treatment of disease and the maintenance of a high quality life.

In an age where it is all too common to find expensive, unnatural treatments for ailments, Dr. Vic Barker has taken us back to nature with chapter after chapter of probing, thought provoking, simple remedies that are based on a study of the evolutionary pathways down which our ancestors have trodden. Why indeed should we fit an articial joint if we can apply natural laws to reverse man made deformities.

My work with fitness centres throughout New Zealand and Australia requires me to constantly seek out new information in the field of physical conditioning and muscular skeletal disorders. In twenty years of research I have never before come across a book so thought provoking as this. Dr. Barker's treatment of osteoarthritis is so simple, logical and effective that I wonder why the medical world has not used it extensively before now. Too often it seems we are so obsessed with the concepts of pharmaceutical remedies or corrective surgery that we do not seek out natural remedies.

Dr. Barker's description of the back and its use together with his treatment for back pain and the prevention of back injury must surely stimulate minds in both medical circles and industry as well as relieving suffering.

This book dares to challenge the accepted theories of muscle contraction and cell division and in doing so opens up new vistas for the treatment of such diseases as cancer. In his description of 'Electric Man' Dr. Barker goes so far as to suggest a method by which the brain is recharged. While admitting these concepts as theories, he establishes himself as an original thinker of some magnitude.

Posture Makes Pefect, with literally dozens of original ideas for the natural treatment of disease and disorders, demonstrates to me that man's mind applied to problems logically can still unlock an unlimited number of doors and give us all a new look at future possibilities. It is a must for the family library and should be essential reading for anyone who is connected with the treatment of human disorders.

To those suffering from osteoarthritis and back pain I recommend that you read it carefully, as it may offer you a new 'painfree' life.

Les Mills M.B.E.

CHAPTER ONE
PHILOSOPHY & FRAMEWORK

"Where is the wisdom we have lost in knowledge,
where is the knowledge we have lost in information?"

T. S. Eliot (1888-1965)

In the rush for technology in medicine, as in other fields, basic principles and simple facts have often been neglected. Fifty years before they were put into practice, the simple techniques which made the treatment of paraplegics so successful, were well known but not applied. It is amazing that a subject so vitally important as the framework to which the rest of the body is attached, and on which it is dependent, has triggered so little interest over the years.

POSTURE IS THAT FRAMEWORK

Over millions of years nature has evolved perfect posture but the architecture is immaculate only when it is not distorted by man's stupidity and arrogance. Imagine designing in great detail the electrical wiring, plumbing, fittings and furnishings of an upright skyscraper, and then putting them all in a building that tilted and curved. Nothing would fit properly or work efficiently. If you built a car with a perfect engine, wonderful suspension, first class braking, and engaged the world's top driver but bent the chassis out of line, the driver's experience and skill would count for little.

The human framework *must* be correctly aligned and this can be achieved by a few simple home exercises. As long as there is no permanent deformity such as a withered arm or shortened leg, your body can be moulded into shape, just as the sculptor moulds a form from clay. The sculptor, however, only achieves motionless beauty and proportion. You can achieve efficiency of movement, reduction in fatigue, and freedom from joint pains, as well as greater beauty of appearance. Once good posture has been obtained, it will remain that way with only moderate effort or maintenance required.

Living in a complicated world of technology we often lose sight of the simple answer. A baboon's heart is put into a man; joints of metal and

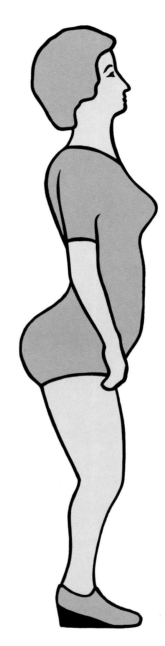

Figure 1-1. Our framework is as important as that of any skyscraper.

plastic are inserted to replace diseased joints. Exciting and expensive achievements! Putting a man's heart into a baboon would be even more newsworthy! But when exercise is proposed to treat the same medical conditions, funds and facilities are very difficult to obtain. Exercise is not

Would you be prepared to sleep in a hotel like this?

enough to pay $200 once for exercise equipment, so that the patient may have no further need of the drugs?

There are great buildings, skilled doctors and nurses, ancillary workers and sophisticated electronic devices specialising in heart surgery, but if the doctor advocates exercise therapy there is little back-up. If the latest and most complex technology, highly trained staff, and great wealth is required to help people, then let us obtain these needs, but *when simple methods are the solution, let us not dismiss them merely because they do not require vast resources.*

Neither the doctor nor the patient should be hemmed in by technology that can prevent wise decisions. Knowledge upon which these decisions are based should include the mundane as well as the exciting, the cheap as well as the costly, the mind as well as the body, the simple as well as the complicated. *Evolution is interested only in whether a thing works.* It does not care how complicated or how costly it is or how cheap and simple or whether or not it is dramatic or of interest to the media.

When a person seeks help for osteo-arthritis, he or she is normally told that it has been caused by wear and tear, a statement seemingly based on medical folklore rather than fact. Myths can originate as mere suggestions of the eminent which then become accepted as facts. Folklore and fact become mixed, until it is impossible to tell the difference between them. Eminence, whether in science, culture, or any other field, brings grave responsibility, for the word of the eminent is taken as gospel. Although there have been many eminent assumptions made on the basis of wear and tear, so many of the very active, from all walks of life, have healthy joints in old age, that this accepted cause must now be questioned.

The osteoarthritis patient is normally told to reduce the wearing and tearing, or to resort to expensive spare parts surgery, or to "learn to live with it". No more despondent, depressing, hopeless advice could be given than to "learn to live with" a painful medical problem. It leaves the patient with the choice between suicide and existing with constant gnawing pain. To really live with it is impossible. Following this forlorn advice, the poor victim is compared with a worn out tyre. He is often told that if he uses himself sparingly, the process will take longer, but he will still wear out. *How cheerless, how lifeless, how untrue!*

exciting, expensive or newsworthy. A physician may prescribe drugs costing fifty dollars a month and in many countries the government foots the bill for as long as the physician says the patient needs them. For osteo-arthritis this can be many years. But which government is forward thinking

CHAPTER TWO

THE LIVING MACHINE

"Nature, to be commanded, must be obeyed."

Sir Francis Bacon (1561-1626)

It is impossible to compare living material and man-made material. Although both the living and the non-living obey the laws of physics, in all its branches without variation or tolerance, their properties and their behaviour are exactly opposite. The tread on a car tyre wears out with use while the tread on the human foot thickens with use. A city teenager's feet are so tender from the constant use of shoes, that he can hardly walk barefoot on smooth concrete, while a Fijian teenager can run barefoot and carefree on a path of shingle. His feet have acquired a thick tread of protective skin, and so would the city teenager's feet, given the same lifestyle.

With stimulus, this thickening of the skin will occur even in the aged. This is growth; this is nature; this is living material reacting to use. Let us be thankful, and use these wonderful properties which the living machine provides. If we take steps to prevent corrosion from the atmosphere, the effect of disuse upon an inorganic (man-made) machine is nil and a car will start immediately, after 50 years lack of use, provided a new battery is installed and fresh water, petrol and oil are provided. On the other hand, if a living machine is not used the parts begin to waste away (disuse atrophy or degeneration). With repeated and constant use, man-made machines are damaged by wear and tear. Conversely, in the living machine, improved strength, efficiency and performance is the response to such stimuli.

The living machine has an ability to repair itself; the ability to revert to its normal structure and function, when outside influences or incorrect functions have been removed or discontinued. If a shoe rubs on a heel a blister results. If the rubbing ceases, the body will repair the damage and normal skin will reappear. No such property can be claimed for man-made machinery which has no ability to repair itself and revert to its undamaged prior state.

Figure 2-1. *With wear on a tyre, the tread gets thinner.*

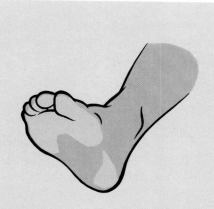

Figure 2-2. *With wear on a foot, the tread gets thicker.*

Figure 2-3. *"As good as new, only 5000 Kms on the clock!" The less it has been used, the better the condition.*

Figure 2-4. *"Stuffed! Need more exercise." The more used the better the condition.*

Figure 2-5. *"The back axle is broken. A new one is needed.*

Figure 2-6. *"Your ankle is broken. It will take six weeks to repair ITSELF."*

	Inorganic Machine (Man-made Machine)	Organic Machine (Living Machine)
Repeated Use	Wear and damage	Improved strength, efficiency and performance in response to stimulus.
Disuse	No effect (provided atmospheric corrosion is discounted.)	Disuse atrophy and degeneration (wasting)
Self Repair	Nil	Ability to repair. Ability to revert to normal structure and function, when outside influences or incorrect functions have been removed or discontinued.

The other vital difference between living and man-made machines is in their response to surgery. For the living machine, there is firstly the danger of the general anaesthetic that is required for the operation. Then there are the dangers inherent in the surgery itself. During the recovery period, there is the risk of a clot forming in a leg vein, becoming dislodged, and travelling to the lung. This misfortune is serious and often fatal. Chest infections such as pneumonia and pleurisy are possible complications. There is also the chance of infection, and when the surgery has been performed on a joint, this infection may occur within a few days, or it may not appear for several weeks.

There may be a breakdown of the surgery itself or a rejection of a spare part by the body. Finally the necessary inactivity that surgery demands results in a loss of general fitness, and degeneration from disuse in other joints and body tissues. None of these complications arise with a man-made machine, provided the repair is done in a logical and knowledgeable sequence.

There is however, no logic in replacing the burnt out valves in your car, without replacing the worn out carburettor which produced them. There is also little point in replacing a man's heart if he returns to the physically inactive life which caused the breakdown of his first heart. And there is little logic if, after a hip joint has been replaced, no attempt is made to correct the poor posture that caused the arthritis in the first place, and may produce further arthritis in other weight bearing joints.

> **Wolff's Law of Physiology, written in 1868 states that "every change in the function of a bone is followed by certain definite changes in its internal architecture and its external confirmation".**

There is enough evidence to extend this principle to all living material, and claim that the structures of the various tissues of the body change according to the functions they are given. When these changes vary from evolutionary design they produce structural deformities. Unwittingly, old time weightlifters were using Wolff's Law in their advice "use it, or lose it". If you do not use your muscles, they will waste away.

Man-made machines cannot change their structure, or the materials from which they are made, in order to adapt to changes in the way they are used. But, for man it is logical to assume that Wolff's Law will apply in reverse. If we correct the function of a human machine, the structure will correct itself.

Examples:

Damaged Structure	Logical Correction	Illogical Method
Muscle	Correct exercise	Male hormones Pituitary hormones
Circulation	Exercise	Surgery, drugs Activity restriction
Joints	Correct posture and alignment	Activity restriction

Bearing these facts in mind, the following chapters examine the multitude of myths that exist concerning our backs, our joints and our evolution into the upright posture.

CHAPTER THREE
THE EVOLUTION OF TWO-LEGGED POSTURE

*"His boneless worm-like ancestors would be amazed
At the upright position, the breasts, the four-chambered heart."*

W.H Auden (1907-1973)

Because disabilities of the back have reached such mammoth proportions in affluent and industrialised societies, it is now claimed that our backs have not been correctly adapted to the upright posture, and were not correctly designed for weight bearing. Why do we blame evolution? This extremely common problem is almost entirely confined to affluent, physically inactive societies; to machine orientated societies; to societies in which much of the day is spent in a chair; to societies in which everyone wears heels on their shoes. These so called 'advances' are the villains not evolution!

> **It is a medical myth that two-legged posture is inherently weak. Actually, it is four-legged posture which is the weaker. Man's column-like structure ensures that the weight is directly and efficiently borne by the skeleton.**

In four-legged animals, the weight is not directly supported, producing an inherent weakness in the middle of the spine, and requiring a suspensory ligament (like a cantilever bridge) from head to tail, to prevent collapse.

Figure 3-1. Man's column-like structure is ideally suited to bear his weight.

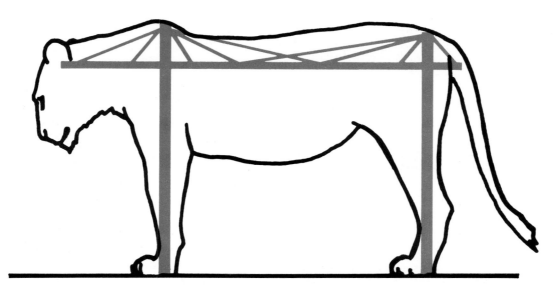

Figure 3-2. The four-legged animal has an inherent weakness in the middle of the spine, which has to be counteracted by a cantilever system as for example, in a cantilever bridge.

If the Asian can work in his rice fields many hours each day until old age, bending at the hip with his knees almost straight, why is the European advised that he must bend at the knees before daring to pick up anything from the floor? If an African can run miles in bare feet without injuring his Achilles tendon, why must an athlete wear heels on his running shoes to protect this tendon? Nature is not to blame. Evolution has done its job beautifully in adapting our spines and our feet to the upright posture. It has had millions of years for research and modification. However, in 150 years, industrialisation has changed our entire way of life. This has been too tiny a moment in time for evolution to adapt to these great changes, and so we must either pay the price in pain and suffering or modify our way of life to fit evolution's design.

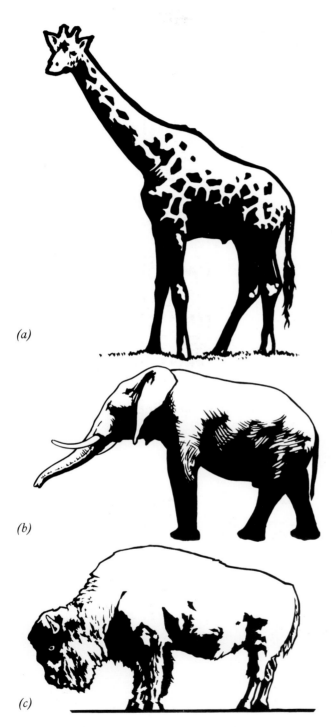

(a)

(b)

(c)

Figure 3-3. In some animals, the spine appears to be sloped to counterbalance the heavy weight of the front of the animal, such as the long neck of the giraffe (a); or the heavy head and trunk of the elephant (b); or the heavy chest of the bison (c).

Figure 3-4. Although its main purpose is to increase the surface area for cooling, the reverse curve of the neck of the camel would seem to counterbalance the hump.

Figure 3-5. The tail of the kangeroo, besides aiding in propulsion would appear to balance the weight of the trunk when it is squatting on its hindquarters.

As well as modifying our spines, evolution has had to adapt our circulation to suit an upright posture. Most of our four-legged ancestors had their hearts and their brains on approximately the same level. This meant that the heart needed to produce very little pressure to maintain circulation to the brain.

In our upright posture, the heart has to pump the thick, viscous blood about 12 inches upwards to the brain.

Evolution has solved this problem efficiently and adequately, even though it would appear to be a much more difficult engineering feat than modifying the spine. So why doubt that the spine has made adequate modifications over the same time span?

Man is a primate, a mammal with advanced brain and eyesight, and a body adapted for both tree and ground living. The early primates lived in the trees of the great forests, surviving on a diet that consisted mainly of leaves, fruits and nuts, supplemented by birds and birds' eggs, lizards, and grubs. To obtain their food, they developed delicate movements of the hands and a strong grip in all four legs for manoeuverability in the trees. Highly developed vision was necessary, in order to pluck accurately at their food, and move from branch to branch at great speed. Three dimen-

Figure 3-6. Heart and brain on same level — requiring only a small pumping pressure.

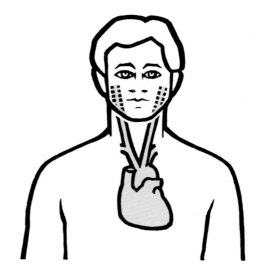

Figure 3-7. Brain above heart — requires strong pump to maintain pressure.

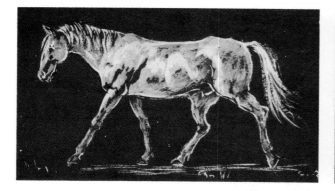

Figure 3-8. When eyes are on either side of the head, they cannot focus on the same object together.

Figure 3-9. With eyes to the front for binocular vision, and under the large brain, the neck has to be bent backwards to see in front.

sional vision, which gave an accurate assessment of the distance of objects was vital. This binocular vision only occurs when both eyes can look at the same object together. The eyes, therefore, had to be brought to the front of the face.

In both monkey and in man, evolution set the eyes in front of the face, and below the comparatively large brain. This meant that our ancestors had to tilt their heads backwards in order to see to the front, — producing the first backward curve of the spine, that of the neck.

In their tree dwelling life, monkeys are frequently in the upright posture. They are vertical when they are climbing the trunk, they hang verti-

cally when they swing from bough to bough, and they are vertical again when they leap in a frog-like action to a higher bough. They sleep in the trees for safety against predators and some even sleep vertically, developing large pads on their buttocks for the very purpose of sitting for hours on a narrow, hard branch. Moreover, their limbs are invariably bent at both the elbow and the knee in order to keep the weight close to the tree trunk and reduce effort.

This flexed position of the limbs occurs also for the attainment of maximum power in tree hopping. Actually, it is only momentarily, when leaping from tree to tree, that we ever find these animals completely upright, i.e. with straightened legs.

Nevertheless, many of these animals can stand on two legs on the ground and can even run in the upright posture but when doing so, neither the hip joint nor the knee joint is fully straightened. These monkeys pull themselves erect by using the comparatively undeveloped muscles of the buttocks. If the spine were one single rigid bone, the manoeuvre would be simple and crane like and if this rigid spine had been developed our back troubles would be almost non-existent. However, in all animals with a back bone, the spine is not one rigid bone, but a series of small bones which hinge on each other. This arrangement has developed to give great flexibility to the spine, so that it can bend sideways, backwards and forwards, and rotate itself. Some of the strength of the spine has been sacrificed in order to produce flexibility.

Figure 3-10. Monkeys keep their limbs bent in order to keep their weight close to the tree.

Figure 3-11. In tree hopping the monkey nearly becomes upright.

The buttock muscles, aided by the hamstrings, raise the lower part of the spine, (the sacrum). Along the back of the bones which make up the spine (the vertebrae) is a series of small muscles — the erectors — which contract to bring the rest of the spine upright. Their action produces a curve in the small of the back.

Thus, coming to the upright position, our ancestors had to develop backward curves in the region of the neck and in the small of the back. If they had not done so, their eyes would have pointed straight down and all their weight would have been on the front of their feet, causing them to over balance forward. In monkeys and apes, who spend little time standing, and never in the upright position with the knees locked, these curves and the muscles which produce them, are poorly developed. But, in man, there are obvious hollows in the neck and the small of the back, with bulges in the buttock and shoulder regions. These curves must counter-balance each other, or we would tend to topple backwards or forwards.

Our nearest ancestors, the great apes, i.e. the gibbon, the orang utan, the chimpanzee and the

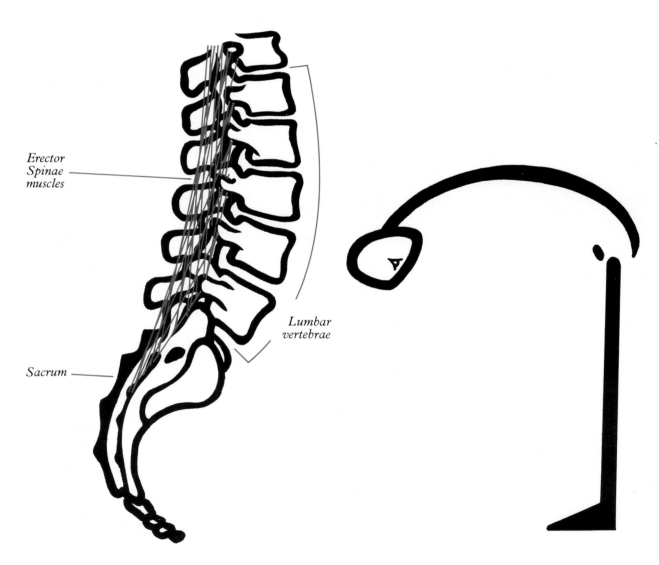

Erector Spinae muscles

Lumbar vertebrae

Sacrum

Figure 3-12. The contraction of the erectors, in pulling man upright, produces the lumbar curve.

Figure 3-13. The upright posture, if we had not developed any backward curves in our spine.

gorilla use all four legs for movement on the ground. However they have long arms in comparison with their legs, — so that they are approaching the vertical, in this type of locomotion. But, the foot of the great apes is still very much a grasping instrument, and is not well designed for walking or bearing the weight of the body. So it has been modified in man. The big toe, instead of pointing sideways like the thumb, is brought into line with the other toes. The big toe of man provides propulsive force in his bipedal striding gait, and his driving running action. The

other toes have become shortened. The heel bone has become enlarged to give a firm base for weight-bearing and greater leverage for the Achilles tendon in walking and running. Only when man runs do his feet show that they originated from a structure designed for grasping. For, as he runs his foot assumes a grasping shape, momentarily, and his weight is borne on the outer side of the foot as is the weight of the great apes in their bipedal locomotion.

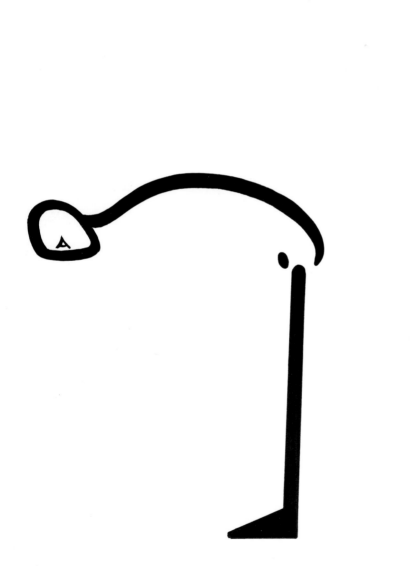

Figure 3-14. The upright posture, if we had only developed a backward curve in our neck.

Figure 3-15. Upright posture, with two backward curves of the spine, at the neck and the small of the back, producing normal upright posture.

CHAPTER FOUR
NORMAL POSTURE

*"Wonders are many,
and nothing is more wonderful than man."*

Sophocles (495-406 B.C.)

Having seen how the curves of the spine have evolved in the human species, let us examine how they develop in the individual.

In the short space of a few weeks, the human foetus, as it grows in the womb, passes through a cycle that roughly parallels the whole of human evolution, from fish to homosapiens. However, the baby does not develop its upright posture until after it has been born. It is completed only when the child has learned to stand and walk. As the baby lies curled in the womb, there is one continuous C-shaped curve to the spine.

Figure 4-1. Posture in womb — C-shaped curve.

When lying in a cot, the baby first experiences the force of gravity, and with ample room to try out muscles, attempts to straighten the back. Nevertheless, the C-shaped curve persists for several months after birth. It is only when the baby sits up that the first curve appears, that of the neck (cervical). For, if the spine kept its C-shape, the heavy head would hang down and vision would be restricted.

By pulling, the relatively heavy head back into the upright position, its weight is brought over the centre of gravity, and at the same time the eyes are pulled up, so that they can look to the front. It is interesting to note that the legs are spreadeagled to give a wide triangular base for support, since balance is, as yet, poorly developed.

Figure 4-3. Posture sitting — cervical (neck) curve present.

The same principles apply when learning to stand up, and to walk. If a baby possessed only a backward curve in the neck, the trunk would hang forward and the eyes would again be looking straight downwards.

Figure 4-2. Posture sitting — before neck curve develops.

Figure 4-4. Baby standing — if no lumbar curve had developed.

To reach the upright position the baby develops a curve in the small of the back, the lumbar curve. This brings the head over the centre of

Figure 4-5. Baby standing — all curves normal.

gravity, the range of visibility is increased and the body weight can be supported by the bony frame. The only parts of the original C-curve which remains are in the chest and buttock regions.

All these curves are marked in babyhood, when the spine is small, but tend to decrease as the spine grows. They are most noticeable in the duck-like waddle of the baby or the dwarf.

Initially, the baby has to learn a sense of balance, so in standing or walking he keeps the feet wide apart, to produce a wide and safe base.

Changes occur in the legs. At birth they are bent at the hip and the knee, but when the child stands erect, these joints are straightened, so as much weight as possible can be borne by the bony frame, and not by muscular effort.

Finally, normal posture is dependent upon the correct positioning of the head upon the spine. This occurs when the head and eyes are looking straight ahead. If the head is tilted either forwards and downwards, or backwards and upwards, the curves in the spine must alter, in order to maintain the centre of gravity in its correct position. The position of the head is vital to good posture.

Looking from the back, one would expect the spine to be straight, with no curves. But this happens only in the ambidextrous. In the right-handed person, the muscles on the right become stronger and the muscle tone greater so the upper spine is tilted to the right. There is a counter-balancing tilt in the opposite direction lower down the spine. Often, the pelvis is also tilted. This sideways curvature of the spine is often marked in tennis players. Men's tailors will confirm that in the right-handed, the right shoulder may be as much as four inches lower than the left, and that the testicles will hang towards the left rather than the right thigh.

The bones of the spine are called vertebrae, and there are seven in the neck (cervical); twelve in the chest region (thoracic); five in the small of the back (lumbar); five fused together into one bone in the regions of the pelvis (the sacrum); and finally, a rudimentary tail consisting of four small vertebrae fused together (the coccyx).

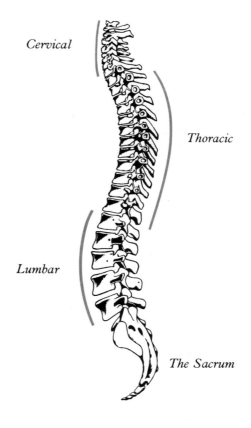

Figure 4-6. The bones of the spine (side view).

Although there are variations throughout the length of the spine so that no two are the same, a typical vertebra consists of a solid "body" in the front, with a ring of bone (the vertebral arch) attached to the back of it. This ring has bony projections jutting out the back, for attaching muscles or for articulating with adjacent vertebrae. The spinal cord which carries the nerves to the various parts of the body passes down the spine through the holes (foramina) which these bony rings make.

stay attached. The common term — a "slipped disc" is a misnomer. A disc cannot slip in and out of place. However, the outer ring may split, allowing some of the jelly like central portion to bulge through. This often requires surgery to repair.

Very little movement can take place between adjacent vertebrae, but the combined effect of movement of the spine is considerable, backwards, forwards, sideways, and twisting. The movements of the spine can also be very complicated, with combinations of these movements acting at the

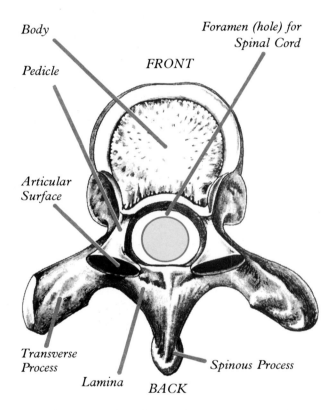

Figure 4-7. A typical lumbar vertebra, seen from above.

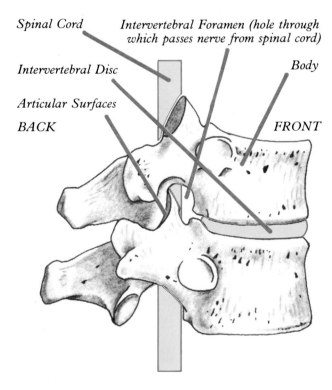

Figure 4-8. Two consecutive lumbar vertebrae, seen from the side.

Between the bodies of adjacent vertebrae is the intervertebral disc — the notorious disc. This acts as a cushion and is made up of a really tough outer ring and a soft, almost jelly-like central portion. The core of this central portion is fluid and therefore cannot be compressed, so it acts as a fulcrum about which the vertebrae can rotate. The tough, fibrous ring is so firmly attached to the vertebrae that the bones may break and yet the disc may still

same time, so there is a complicated system of muscles which can control these movements. To bend the spine backwards, there is a series of muscles running down the back. To bend the spine to one side or rotate it, only the muscles on one side are activated. In a well muscled person they can be seen as thick masses running down either side of the spine and give the clue to ability in most sports, for they are the very seat of strength. In no other

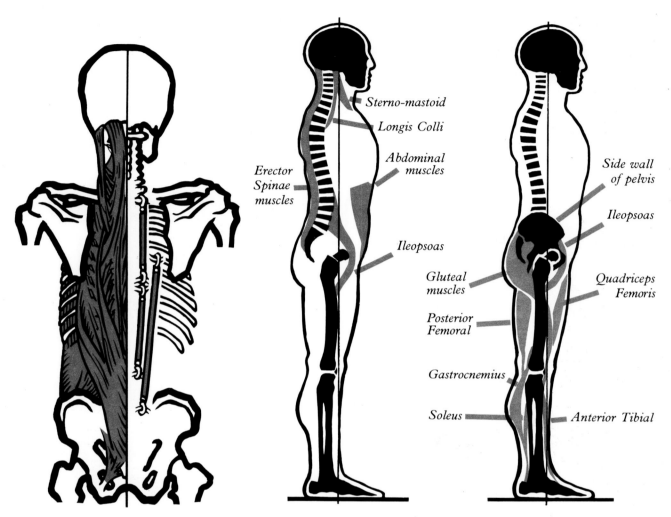

Figure 4-9. Diagram to show how tension in muscles resemble elastic springs. Left side — dissected muscles, Right side — elastic springs.

Figure 4-10. Side view showing flexor muscles of spine balancing extensor muscles.

Figure 4-11. Side view showing muscular balance around hip, knee and ankle.

part of the body do muscles, when dissected, look so much like the strands of elastic chest expanders. It is the tension in these muscles that holds the spine in its correct position.

The tension in the back muscles must be balanced by the tension of muscles in the front of the body which bends the spine forward. These are the muscles in the front of the neck, the abdominal muscles, and the muscles which bend the thigh forward at the hip joint.

The positioning of the head on the spine depends also on the tension of the neck muscles. The alignment of the bones at the hip, knee, and ankle

joints which are held by the tone of the muscles which act on these joints, is also vital to correct posture.

> **It is absolutely essential to consider all the joints that bear weight as one Weight Bearing System.**

It is because each joint (whether of the spine, hips, knees or ankles) has invariably been studied

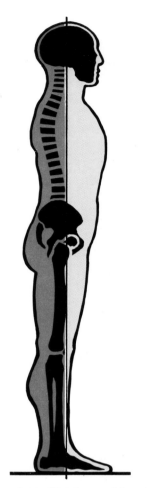

Figure 4-12. Weight on either side of the vertical line through the centre of gravity must balance.

Figure 4-13. End joint of finger, straight and bent.

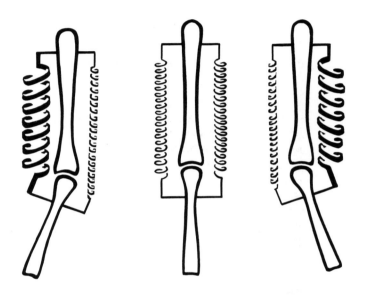

Figure 4-14. Showing muscle tone as the tension in a coiled spring. The joint bends towards the stronger spring (tension).

in isolation, that correction has been so unsuccessful.

If you draw a line from just in front of the ankle to the top of the head (the line through the centre of gravity), then any curvature on one side of this line must be balanced by an equal curvature on the other side. If one curvature becomes exaggerated, or flattened then another equally exaggerated or flattened curve develops on the other side of the line.

The counter balancing curves of the spine are kept in alignment by muscle tone, in the same way as every other joint in the body. If the muscle which bends the finger, for instance, is stronger than the one which straightens it, the finger will tend to stay bent, and it may, in time, become impossible to straighten.

Chapter 12 will show how this bent position produces the knob shape and pain of osteoarthritis.

Muscle tone is like the tension in a coiled spring. The joint bends toward the stronger spring.

Using this simple principle of relative muscle tone, any normal joint in the body can be adjusted so that it maintains its correct position, and its flexibility can be increased to its full range.

The correction of poor posture is very simple. One has only to strengthen the weak muscles, and the joints in the spine will be gradually pulled back into their correct positions. Of course, posture is

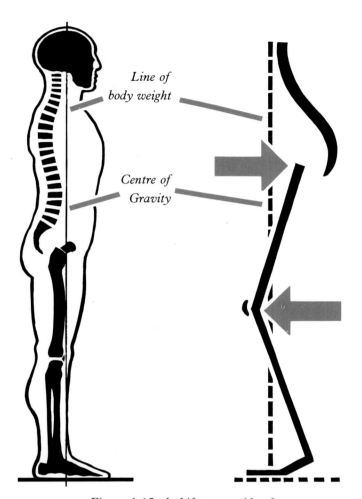

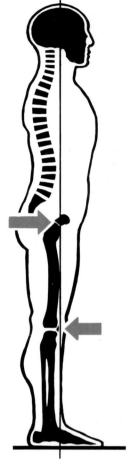

Figure 4-15. A shift to one side of the vertical line through the centre of gravity, must be counter-balanced by an equal shift to the opposite side.

Figure 4-16. The body weight assists in locking hip and knee, when standing.

also dependent upon the position of the head and the alignment of the legs and feet, and these can be treated by the same principle. If the knees are bent, weight is shifted forward. If the hip is bent weight is shifted back. The head can be tilted backwards or forwards or to either side. All need a counter-balancing shift of weight to keep the centre of gravity in the right place.

In normal posture, the line of the bodyweight passes behind the hip joint, so the bodyweight tends to push the joint forward, straightening it. The same vertical line passes in front of the knee joint so that the bodyweight locks this joint backwards to straighten it. Thus bending and sagging at both these joints is avoided.

The mechanics of posture consist of balances and counter-balances. These keep the human frame upright and are the keys to grace and beauty. Any variation in the angles at any of the weight bearing joints (spine, hip, knee, or ankle) displaces the bodyweight, and an equal and opposite displacement in another weight bearing joint must take place in order to maintain the line of the bodyweight. The angle at any joint varies with the tension of the muscles acting on either side of it.

When all these factors work in harmony with each other, as they are meant to, the body is in harmony with itself, and improved health and performance are the results.

CHAPTER FIVE

POSTURE OF THE SHORT
AND POSTURE OF THE TALL

*"While the tall maid is stooping,
the little one has swept the house."*

Italian Proverb.

People vary from the normal in their posture. There are a number of reasons for this, the main being variations in height.

Those who are below average height spend much of their lives looking upwards, whereas those above normal height spend their time looking downwards. When shorty is standing on the street corner talking, he finds he is looking upwards. When he goes to work, whether in the office or the factory, everything is too high for him. When he goes to lunch the chair is too high and the table too high. When he drives home he has to pad his car seat in order to peer through the windscreen. At home, the sink bench, the cupboards, and all the furniture are too high.

Lofty's problems are the opposite, but just as great. He finds that everything is too low and too small, and he is forever trying to lower himself as he tries to fit in with the people, the surroundings, and the equipment he is forced to use.

The posture of both the tall and the short suffer as they try to fit themselves into an environment designed for the average person. The short person's head tilts back as he looks up, and this increases the neck curve in the spine, with a compensatory decrease in the normal curve of the chest region. He also arches the small of his back which increases the lumbar curve. This makes his bottom stick out, and his thigh bones (femurs) angle forward, instead of straight down. This, in turn, pushes his knees forward tilting his lower legs forward from the ankle joints. The head, neck, and buttocks, in moving back, have pushed the belly, thighs and knees forward as a counterbalance. The hip and knee joints have unlocked and the pressure on these joints has been reduced, (see Chapter 12). His weight is no longer fully borne by thigh and lower leg bones, but has to be partially supported by muscular effort.

This type of posture has all the features of the duck, including the typical waddle and can be aggressive in appearance. It is more common and more obvious in women who are on average, shorter, and whose thigh bones are angled in a different way at the hip to that of the male, (see Chapter 11).

With age these variations become more pronounced, and accumulations of fat in stretched muscles further exaggerate the caricature.

The tall person's head tilts forward to assist the downward gaze, and the curvature in the region of

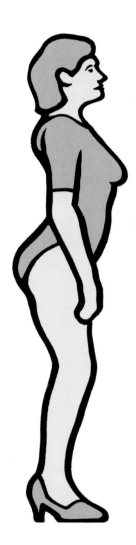

Figure 5-1. The posture frequently developed by the short, because of the need to look up.

the chest is increased producing a hump back. The shoulders droop forward, producing round shoulders and a pigeon chest. The knees and ankles bend, in order to reduce height and to counterbalance the hump back. And in order to keep the centre of gravity over the feet, the curves in the small of the back and the buttock region are flattened. The total effect is of head, shoulders, arms and legs thrown forward, whilst the trunk itself is moved backward to maintain balance. The stoop of the tall slim person is also often exaggerated because of self consciousness.

Many other factors besides height affect posture. Type of work, sporting activities, personality, heredity and disease are major influences. Some short people have round shoulders and a stoop, and some tall people have arched, sway backs, but these people are exceptions, as the two basic postures are most often related directly to the height of the person.

Since the posture of the short is characterised by an increase in the curve of the small of the back (lumbar lordosis), they are known technically as lordotic types. The increased curve in the chest region, which is characteristic of the tall person (thoracic kyphosis) results in them being called kyphotic types.

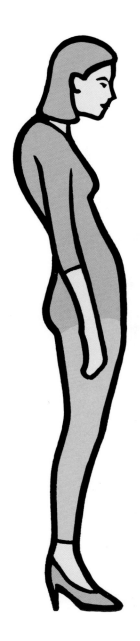

Figure 5-2. The posture frequently developed by the tall, because of the need to look down.

Treatment
See Chapter 21 for:
(A) **Essential Exercises 1, 2, 3.**
(B) **Recommended Exercises 5-10.**
(C) **Recommended Exercise for round shouldered 4.**
(D) **Recommended Exercises for sway-backed 22, 23.**

CHAPTER SIX

POSTURE AND FOOTWEAR

"Old friends are best.
King James used to call for his old shoes;
they were the easiest for his feet."

John Selden (1584-1654)

There is no doubt that many of our posture problems arise because of our inherited height. We have no say in how tall we are, but we do have a say in the other main cause of poor posture. We can choose whether or not to wear shoes with heels.

It is not known for sure when heels on shoes were first used. The Greeks used them to increase their height, and the ancient Persians used them to keep their feet off the burning sand — although walking on sand is made even more difficult by the use of heels. Mongolian and other oriental horsemen used heels to keep their feet in the stirrups.

However, widespread use of heels on footwear did not take place until the end of the sixteenth century. Pattens, a type of slip-on wooden overshoe, were used throughout the sixteenth century. They raised the feet several inches to negotiate the garbage in the filthy streets, but they raised the heel and sole evenly, so that the feet were on the same level. Stilt shoes, known as chopines, were a female Venetian fashion which lasted twenty years. They raised the wearer as high as eighteen inches above the ground, on a true wedge heel which was up to two inches higher than the sole. In crossing streets littered with garbage, they were invaluable, es-pecially when wearing the awkward farthingale. But walking was so difficult that an assistant on either side was often required.

The fashion for heels probably started in France and spread to England about 1560, where it was at first greeted with a certain amount of derision and satire. Swollen calves, sprained and broken ankles, and sore toes were so frequent that the fashion faded, for men, after a few years and after about twenty years for women. Unfortunately it soon returned popularised by James I, and has been with us ever since. James I was short, with legs distorted by rickets. The royal bootmaker added heels for regal height, and produced what we now know as a cavalier style boot to disguise the King's infirmities.

Men have always preferred a lower heel than women, because of their more active way of life and the differences in their anatomy at the hip joint (chapter 11). However, work patterns are changing, and male dress has become more flamboyant in recent years. Today's young men often wear high heels, and we can predict that males will suffer increasing disfigurement, disease and injury as a result.

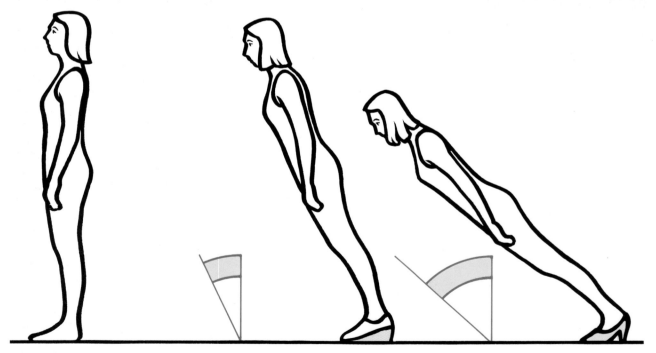

Figure 6-1 (left). Normal posture. Figure 6-2 (centre). Tilt (about 25°) produced by 8 cm. wedge heel, without any correction by the body. Figure 6-3 (right). Tilt (about 45°) produced by 8 cm. heel before any bodily correction.

It took over twenty million years for man to evolve from the four-legged to the upright posture. Heels, a new factor, have been introduced only in the last four hundred years. Heels do not suit the upright posture.

Yet, when things go wrong with our functioning, as the result of this new factor, we have the colossal arrogance to blame evolution and to claim that our upright posture has caused the problems. Shoes, not evolution, cause the problems. They are a retrograde step in evolution — pushing us back towards the four-legged posture. It is no coincidence that osteo-arthritis has been labelled the scourge of Anglo-Saxon hips. Shoes with heels originated in Europe, and, unfortunately, the fashion has spread almost world wide.

The tilt produced by an eight centimetre heel on a 1.80 metre person is approximately 25 degrees if the shoe is of a wedge type. A normal last increases this tilt to 45 degrees. Even with a 2 cm heel, the height of the average man's shoe, the tilt

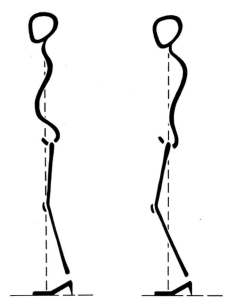

Figure 6-4 (above left). Modifications made by sway-backed (lordotic) types to compensate for raised heels.

Figure 6-5 (above right). Modifications made by round-shouldered (kyphotic) types to compensate for raised heels.

is 12 degrees.

Of course, we cannot possibly walk around like this, so we have to make considerable correction to avoid toppling over and to bring our frame into something resembling upright posture. In doing this, the spine and the weight-bearing joints become distorted and misshapen. It is not surprising that they eventually voice their protest in the pain of osteo-arthritis and many other postural problems. The ankle joint is bent back towards the heel, the knees and hips bend and the spinal curvatures are exaggerated. In the sway back (lordotic) types the lumbar curve becomes increased, whereas the round shouldered (kyphotic) types become even more bent.

Every joint that takes part in bearing the weight of the body has its normal position altered by the wearing of heels, and when the angle at the joint changes, the tendons, ligaments, capsules and muscles working on this joint have their normal length and, therefore, their mode of action changed. Furthermore, the pressure on the bones and on the cartilage lining the surface of the bones in the joints, is altered.

Shoes have two purposes. They are designed to protect the feet and enhance appearance. Both these functions can be fulfilled without the use of heels.

Most people are aware that raised heels can cause bunions, but this is the extent of their knowledge. Even on their own, unsightly bunions seem to be a large price to pay for fashion. A barefooted woman whose big toe creeps over the top of the second toe, and whose feet are disfigured by bunions loses much of her beauty. However, when the public and the shoe manufacturers become aware of the multitude of disfigurements and pain which result directly from heels, we can expect that a wide variety of fashionable shoes without heels will become available. The material and design of the uppers could be as varied as ever. In fact, because the body weight is not thrust forward on to the ball of the foot and the toes, designs such as pointed toes, square toes, and peep toes are much more comfortable. With heels on shoes, the foot slides to the front of the shoe. Without heels, the weight remains towards the back of the shoe, and a smaller size may be worn. The thickness and design of the bottom of the shoe could have great variation, provided only that flexibility be retained, and the range of colour and material for both bottoms and uppers would be no different from that available in shoes with heels.

As there is some loss of grip on slippery or sloping surfaces, when heels are removed from footwear, it is advisable to add some form of non-slip surface to the whole length of the underside of the shoe.

It should be noted that shoes with the sole higher than the heel, making the person walk uphill, cause as many complications as shoes with heels.

If we were meant to have raised heels, the raise would be within our bone structure, not in our shoes. Hoofed animals walk on their toe-nails. How much betting money would be placed on a horse who had the backs of his shoes raised by two inches? Other animals walk on their toes. The elephant has brought his true heel bone down to the ground to bear his enormous weight, and this is why his walk is so much like a human's. Man has evolved to have his heel bone on the ground. As we have been foolish and conceited enough to change nature's design by raising our heels, we have had to pay for the damage and distortion in pain and inefficiency.

Short women should realise that raised heels induce buckling of the weight-bearing joints, so that the gain in height from heels is not nearly as great as the height of the heel. With age, this buckling becomes more pronounced, increasing the loss of true height. A flat, hinged true platform shoe — fashionably designed — would increase height without distorting posture, and be very attractive.

Viewed from behind, or in motion, high heels can detract from a woman's beauty and grace. If mirrors showed the back instead of the front view, high heels would lose much of their popularity.

It is usually insisted that a shoe should give good support to the foot. The correct treatment for a sprained ankle is to support it, and, in doing so, to rest it and prevent further injury; the correct treatment for a damaged Achilles tendon is to raise the heel of the shoe, in order to rest the tendon; a broken ankle is rested by encasing it in plaster. But no one would suggest putting the ankles in plaster in order to prevent broken ankles. It is just as illogical to suggest that a shoe should support the foot in order to prevent injury. The shoe that gives support prevents the muscles, ligaments, and tendons from doing their natural job of supporting the foot. This kind of shoe can be dangerous, particu-

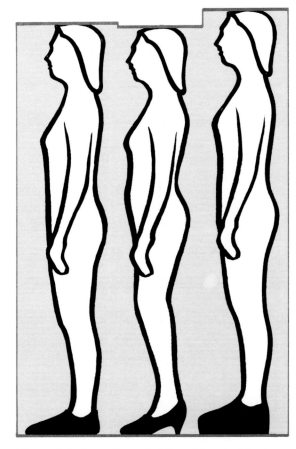

Figure 6-6. Loss of true height with raised heels.

larly in children, and old people.

Evolution has developed the foot into a beautiful shock-absorbing system designed to work efficiently, even over rough, uneven ground. If the foot treads on a large stone, it moulds to the shape of the stone, while the ankle adjusts to the variation, (Figure 6-7).

In countries where many young people go without shoes, it is noticeable that few suffer sprains and breaks. If you are wearing a so-called good, supporting shoe when you tread on a stone, the natural mechanism of shock absorption cannot operate. There is a sudden wrenching of the whole foot and ankle to one side, and a severe sprain or even a broken ankle may result (Figure 6-8).

Flexibility is absolutely essential in the design of shoes. Not only should there be side to side flexibility, but also flexibility in a lengthwise direction. Whether walking or running the human foot rolls naturally from heel to sole to toe, with a

Figure 6-7. Moulding of human heel to uneven ground.

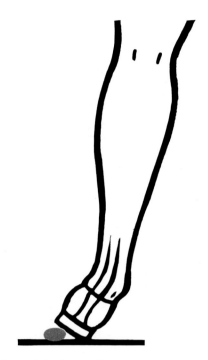

Figure 6-8. A so-called good, supporting shoe permits no moulding on uneven ground. A broken or sprained ankle is a likely result.

strong thrust coming from the sole and particularly the big toe. Some modern shoes are so inflexible that this roll is quite impossible. The walking motion then becomes a matter of placing one foot in front of the other, with the foot playing very little part in the activity.

This is especially true of shoes for toddlers, most of which are very difficult to bend in either direction. Some, with plastic soles are extremely rigid, and of course, the heels have no flexibility. With a child's comparatively weak muscles, flexibility in shoe design is especially important. Otherwise, the beautiful flexibility of a child's feet and ankles is soon reduced. It is not surprising that children pull off their shoes at every opportunity.

Where possible, it is better to let a child run barefoot or wear flat sandals. His feet are well insulated by fat, and circulation is good at this age, so the cold is not usually a problem. In bare feet, the child will develop a natural, free gait with good balance and correct posture and will retain flexibility in feet and ankles, and strength in muscles, tendons and ligaments. The skin will thicken and harden, providing a protective shield which is nature's own form of footwear.

Ballet dancers and Shakespearean actors wear shoes without heels and they achieve a grace and fluid movement that most other people do not have. It is the grace that the black races had before they adopted European shoe fashions. It is the grace the Indian woman has in her sari and flat sandals, a grace that disappears rapidly, when she switches to high heels.

Shoes with heels reduce flexibility, and because the heel gets in the way the normal heel to toe roll of the foot is relatively impossible. The reason why heels on shoes wear out so rapidly is simply because they are there.

The body has developed a perfect shock absorbing system for walking, running and jumping. Besides the rolling action of the foot, there is give at the hip, knee, ankle and smaller joints of the foot. If flexibility is decreased and heels are added to shoes the efficiency of this shock absorbing system is markedly reduced, due to the misalignment of the joints within the system, and the fact that shock absorption at a joint is most effective when the joint is almost straight (Chapter 20).

Besides the postural benefit of wearing flat

Figure 6-9. The effect of heels on the hang and hem of a dress.

shoes, there are a number of other benefits. Flat shoes give relief to those who suffer from bunions, corns, gout and ingrowing toe-nails, because the body weight is taken away from the front of the shoe. For those who suffer from diabetes or other causes of poor circulation to the lower legs and feet, heel-less shoes would seem essential since pressure on toes, toe-nails or blisters on these people may lead to infection or even gangrene.

With high heels the foot angles down and tends to trip on any obstacle. Without heels, the rear of the shoe strikes first and the toes come down on any obstruction with the body weight over the heel. This is of particular advantage to the sightless, the aged and the infirm.

As shoe manufacturers are not presently prepared to give us heel-less fashions we have no alternative but to remove the heels from existing shoes. As soon as I buy a pair of new shoes I have the heels removed and the entire length of the underside of the shoe covered with a thin flexible non slip layer. This is only practicable when shoes have heels no higher than 2 centimetres. Any more than this will cause the toes to turn up too much because the shoes were designed for heels. It is common to mistake the popular wedge shoe for a heel-less shoe. The heel is still there, and even a half a centimetre difference between sole and heel level will distort the posture and the joints considerably. If a tall building was tilted one degree, the doors, windows and carpets would not fit, the lifts would not work, and all the wiring and plumbing would snap. Our bodies modify for such distortion, but eventually, we pay in stiffness, decreased efficiency and pain.

The aim of a shoe is to protect the foot from cuts, bruises and the discomforts of weather. It should fit as closely and be as flexible in all directions as a glove. These are the basic requirements, and fashion designers should be capable of producing a multitude of exotic variations, without deviating from these principles.

CHAPTER SEVEN

POSTURE AND WIDTH
A NEW ANGLE ON OBESITY AND DISEASE

"Jack Sprat could eat no fat,
His wife could eat no lean;"

16th Century Proverb

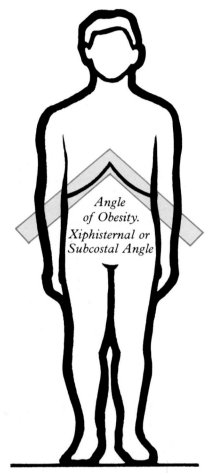

*Angle
of Obesity.
Xiphisternal or
Subcostal Angle*

Figure 7-1. The angle of obesity.

In order to type physiques, I decided to measure the angle (subcostal angle) that the ribs make where they join the bottom of the breast bone (the xiphisternal notch).

I reasoned that if this angle were greater than average, the abdomen would be wide and deep with plenty of room for large digestive organs, whereas the chest would be comparatively small, with a small lung capacity. I considered that these people would have great strength and a tendency to fat, but their stamina would be limited.

If the angle were small, the abdomen would be narrow and shallow with a small digestive system, but the chest would be long with comparatively large lungs. These people would have great stamina, but their strength would be poor, and they would tend to be thin.

I have called this angle *"The Angle of Obesity"*.

In 1973, when working as a school medical officer in England, I measured 143 children attending consecutively for routine medical examination. I chose children because their degree of obesity had not yet been greatly affected by their work patterns or posture. I divided them into five groups, according to their appearance; normal, tending to be fat, fat, tending to be thin, and thin.

The hypothesis was well founded, for nearly all the fat children had angles greater than ninety degrees and nearly all the thin children had angles less than ninety degrees. The average angle was ninety degrees and occurred in 37%. 72% had angles between 80 degrees and 100 degrees. The range was quite startling, varying from 60 to 130 degrees. Below seventy degrees the children were skinny; above 110 degrees they were definitely fat.

It is very easy to measure the angle of obesity. Simply run your fingers down the front of your chest bone until the bone disappears. Mark a cross on your skin there. Now, run your hand down the lower border of your ribs on the left side and mark another cross at the outer edge of this border, approximately where your elbow touches the side of your chest. Do the same on the right side. Now measure the angle between the side crosses and the centre cross with a protractor.

Anybody with an angle greater than 100 degrees is going to fight the battle of the bulge all their life, and above 110 degrees, even the smell of food will put on weight.

The best way for these people to control their weight is either by burning calories with exercise such as running, cycling and skipping (though this is not to their liking and is very hard for them because their stamina is low initially), or by repetitive, explosive type exercise such as weight-training, which may be more to their liking.

This problem is discussed more fully in chapter 9 (Obesity). That obesity is related directly to the size of the digestive system and therefore to the angle of obesity, is confirmed by the success of the intestinal by-pass operation, in which some of the bowel is short-circuited, leaving only a part of the bowel functioning.

Although it helps to reduce weight, this operation is fraught with complications, sometimes continuous and frightful diarrhoea and is "ecologically illogical". In fact, it seems to be a terrible assault upon the body, for a condition which can so readily be treated with a skipping rope.

Another operation which confirms that obesity is directly related to the size of the digestive system, is the gastric by-pass operation in which a large part of the stomach is by-passed. The risks of surgery seem unnecessary. Furthermore, many of these people are not nearly so much overweight as the illogical weight tables dictate (Chapter 9).

Fat people are teased and harrassed unmercifully when they are children. They learn to put on an armour of jovial good humour and wit to cover their misery, or they resort to petulant silence, mixed with grumbling. By the time they have matured, because their large digestive machinery ensures the maximum utilisation of food intake, they have often gained considerable size and strength, but most are already mentally attuned to being dominated. When they apply for work, the odds are stacked against them, because they are seen to be (and sometimes are), clumsy, slow, accident prone, and sweaty. Very fat people may be stared at in the street and avoided in buses, aircraft, or pubs, where they take up too much room.

Yet their angle of obesity, and therefore their obesity, is inherited just as surely as is the colour of their skin. I have seen this over and over again. One great example was of a father and son, both of whom had angles of 130 degrees and both weighted about 127 kilograms.

People with angles of less than eighty degrees will remain thin for most of their lives. From middle age, they may develop a paunch from bad posture but weight will not be gained elsewhere. The only hope such people have of gaining weight, if desired, is to increase muscle size through exercises, such as weight training. (See Chapter 21).

By knowing your angle, you have a better chance of reaching and maintaining your ideal weight. If the angle is known early in life, it will be easier to prevent the problems of obesity. From the outset, a careful check can be made on a baby's weight gain which should be between five and six ounces a week. If the gain is greater than this the baby is being fed too much. This is especially important in breast fed babies, because the amount of food they are getting is unknown. Hunger is a discomfort, not a pain. But colic, which is frequently the result of over feeding, can be agonising. Starving babies rarely cry. They lie listless and still. The judgement of a baby about the wisest quantity of milk to drink is about as accurate

as men with free beer. And the result is the same — indigestion and loose bowels. In fact, the bowels are a good feeding barometer. Over feeding generally produces loose and frequent bowel motions whilst infrequent and hard motions usually denote under feeding.

It is important to note that, on average, a baby's angle of obesity is about ten degrees greater than the corresponding angle in adults. I confirmed this by checking babies attending clinics for routine medical examinations. This finding was to be expected for babies grow more in height than in width. If the parents of babies with wide angles are careful with feeding, they can take pride in knowing that they are preventing obesity and heartbreak for their children in the future. Bonny, bouncing *fat* babies are no reason for pride.

On the other hand, the baby with a narrow angle (less than ninety degrees) will have the opposite feeding needs, and should not be expected to gain weight quickly. These babies will always weigh less than average. They need frequent, small feeds, whether breast or bottle fed. Big feeds may well produce the agony of colic and vomiting or diarrhoea may occur to get rid of the excess food. The result is weight loss rather than weight gain, and a vicious and dangerous circle may result.

The incessant crying of a baby with colic can bring misery to a family and in extreme cases a battering to the baby.

I recall a baby of three months who was brought to me weighing just over eight pounds and actually losing weight. He had already been thoroughly investigated in hospital for failure to gain weight. The angle of obesity was small. By drastically reducing the size of the feeds, the baby was induced to gain, weighing just over eleven pounds at six months. Unfortunately, circumstances took the baby to hospital. A multitude of tests could find no abnormality that would account for the slow progress.

Comparatively large feeds were reintroduced, and weight loss returned. The same thing happened at a second hospital to which the baby was admitted, and death occurred at eighteen months with the baby weighing just over eight pounds. Three years after this sad event, the parents returned to my area, and I was able to confirm that the father had a small angle of obesity and was skinny. By this time they had another baby which matched the mother's normal angle. This baby thrived.

Although the length of a new born baby is carefully recorded, no attempt is made to measure its breadth, yet each measurement is as important as the other. The angle of obesity is a correlation between the two, and more important than either in the assessment of development and in the tendency to illness.

Doctors and Insurance companies would do well to consider this angle as well as measurements of height and weight. With a wide angle, a six foot man may be extremely fit at 95kg, but considerably underweight at 80kg, but with a narrow angle, a six foot man may be fit at 70kg, and grossly overweight at 85kg.

Loaded insurance premiums for many so called overweight are unfairly and unscientifically determined.

THE ANGLE AND DISEASE

It is quite possible that the angle of obesity is relevant to other conditions, because the relative sizes of the abdomen and the chest are controlled by this angle. A narrow angled person has large lungs and great stamina, and can be expected to have a large volume of efficient lung remaining, even if attacked by pneumonia. However, without sufficient exercise, much of the upper part of the lung will be largely unused. This poorly ventilated upper part of the lung is the common site of tuberculosis. It would seem logical to treat TB with exercises to use this part of the lung, as well as treating the disease with modern drugs.

Lung ventilation can be quite great in the narrow angled person, because of the large range of movement possible in the lower ribs and diaphragm.

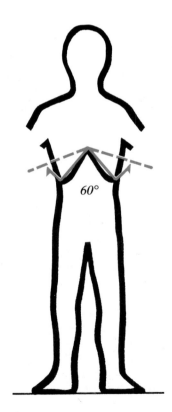

Figure 7-2. Large range of movement of lower ribs in narrow-angled.

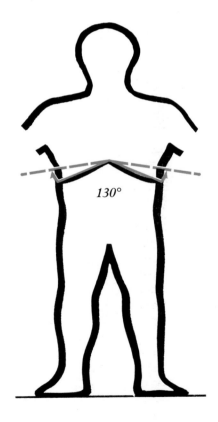

Figure 7-3. Small range of movement of lower ribs in wide-angled.

In the large angled person, the movement of the chest is shallow.

Fat people are considered to be more susceptible to bronchitis and other chest conditions and although unproven, it is the fatness which is blamed. It would seem more logical to look at the wide angle and the small lung volume.

HEART DISEASE

Heart disease is another problem that is said to be related to fatness. In the wide angled person, the heart is likely to be relatively smaller because of the relatively smaller chest cavity.

It tends to lie horizontally so that the main artery from the heart (the aorta) forms a spiral, rather than the "single plane" curve of this artery in the narrow angled. This spiral spins (centrifuges) the blood which has been pumped from the heart under great pressure (a phenomenon which may affect the laying down of cholesterol and other substances in the walls of the blood vessels), causing it to impinge more on one area of the wall of the artery. Research may well establish that this spiralling of the aorta plays a part in the incidence of aortic aneurysm (ballooning of an artery) arteriosclerosis (hardening of the arteries) and angina (reduced blood supply to the heart itself).

THE LIVER

Sited below the chest we find that the liver in the small angled person is generally smaller because the abdominal cavity is smaller. A smaller liver implies greater risk if hepatitis is contracted, or cirrhosis develops.

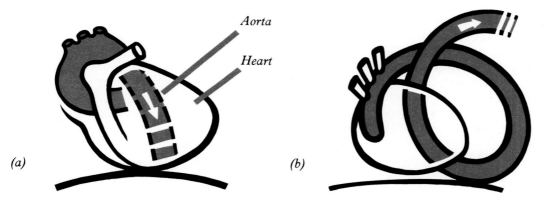

Figure 7-4. (a) In the wide-angled, the heart is comparatively smaller and tends to lie horizontally. (b) The aorta tends to form a spiral.

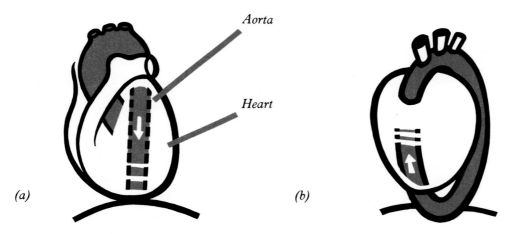

Figure 7-5. (a) In the narrow-angled, the heart is comparatively larger and tends to be vertical. (b) The aorta tends to form a "single plane" curve.

ULCERS

Ulcers are said to be more common in the thin worrying type of person. The problem is how to measure worry. Facial expression and an excess of nervous energy have traditionally given rise to people being labelled as 'worrying types'. However, after 25 years of general practice, I believe that the incidence of ulcers is directly related to the angle of obesity, being greater in those with a narrow angle. These people may look more worried because they have many facial creases but there is no evidence to indicate that they worry more than the fat (wide angle of obesity) whose faces are unlined.

In fact, when a fat person loses weight rapidly, skin creases appear and the face looks less jovial even though the person may be much happier because of the weight loss.

The 'excess of nervous energy' may simply be an abundant supply of physical energy possessed by the narrow angled person who has greater stamina and carries a lighter frame. The ulcer may develop because of a relative excess of acid production and/or concentration in the narrow angled person's smaller stomach or because the valve at the bottom of the stomach (the pylorus) does not function efficiently because the curves of the stomach and the intestines are sharp (since the abdomen is cramped for space).

These sharp curves can delay the acidified food in its passage through the first part of the intestines (the duodenum), or cause digestive juices from the stomach to impinge on the side wall of the duodenum instead of being squirted along the duodenum.

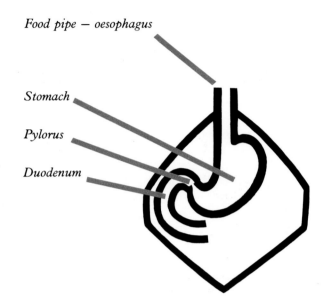

Food pipe — oesophagus

Stomach

Pylorus

Duodenum

Figure 7-6. In the narrow-angled the curves where the duodenum joins the stomach (i.e. at the pyloric valve) are sharp.

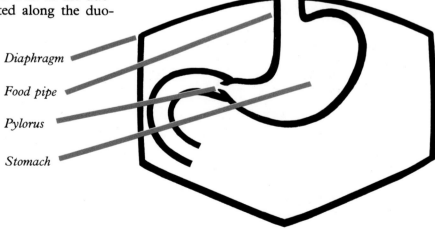

Diaphragm

Food pipe

Pylorus

Stomach

Figure 7-7. In the wide-angled the curves at the junction of the stomach and the doudenum are gradual.

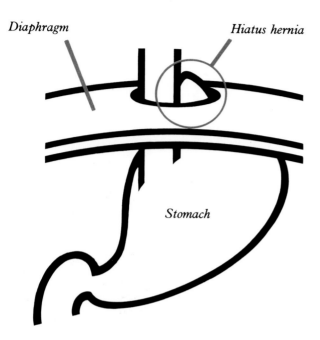

Figure 7-8. An Hiatus Hernia occurs when part of the stomach ruptures through the diaphragm into the chest.

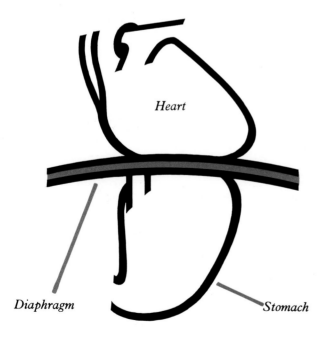

Figure 7-9. In the wide-angled, a large area of the surface of the heart is separated from the stomach only by the diaphragm.

MISCELLANEOUS PROBLEMS

Do the fat get gall bladder disease and gallstones because they are fat, because they have broad angles or because of poor posture (Chapter 17.) The broad angle produces a large liver and possibly an excess of bile, and may produce a greater curve in the tube taking the bile from the liver to the intestines.

Hiatus hernia is a condition in which part of the stomach moves up through the diaphragm into the chest.

It frequently causes pain and discomfort. It is more common in the fat, and appears to be related directly to the angle of obesity. It may be more prevalent in the wide angled because of the flat shape of the dome of the diaphragm. However, the small lung capacity may mean that the diaphragm has to be moved more vigorously, or the short food pipe (oesophagus) may be stretched by the lordotic posture of the wide angled. In paroxysmal tachycardia, the heart, for no obvious reason, suddenly starts to beat at more than twice its normal speed. As suddenly, it returns to normal. Although rarely dangerous, this is a very distressing condition. The heart tries to beat before it is filled with blood, and the person soon becomes short of oxygen, in spite of the fact that breathing is normal. I believe it is more common in the wide angled. The combination of exercise on a full stomach can be a precipitating factor. In some women it may only occur during pregnancy, and it is certainly much more difficult to treat in the pregnant. It is likely that pressure transmitted from the abdomen, through the diaphragm, to the heart is an important factor. In the wide angled a greater surface of the heart is in contact with the diaphragm, affecting the likelihood of this problem.

THE ANGLE OF OBESITY AND SPORT

This is a particularly useful measurement for advising a youngster's aptitude for sport. For sports and events requiring great stamina, an angle of 70° or less is advisable. Where great strength is required an angle of 110° or more would be almost a necessity. For most team sports, the best players are likely to have the normal 90° angle, with stamina usually of more use than strength.

THE ANGLE OF OBESITY AFFECTS POSTURE

The wide angled, with their natural tendency towards strength activities, usually develop powerful backs with an exaggerated curve in the small of the back (lordosis). This curve is counter-balanced by an excessive curve in the sacral region, so the buttocks protrude further than normal.

In people of average height there is unlikely to be any other modification to spinal curves. In short people both the thoracic and cervical curves are likely to be reduced as the result of the necessity to look up.

In tall people, the thoracic and cervical curves will be exaggerated due to the need to look down — the excess curvatures counter-balancing each other to keep the line of the centre of gravity constant.

In such people there is lordosis in the lumbar region and kyphosis in the thoracic region and these people are suffering from kypho-lordosis.

The narrow angled have comparatively weak

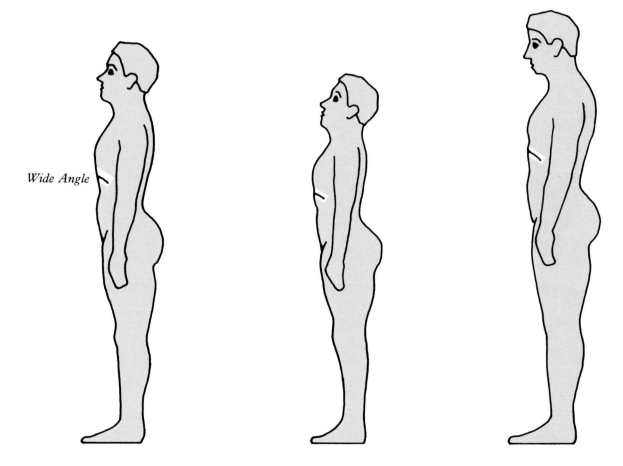

Wide Angle

Figure 7-10. Lordosis (sway-back) is usually associated with a wide-angle of obesity.

Figure 7-11. In short people, with a wide-angle of obesity, the cervical and thoracic curves flatten.

Figure 7-12. In tall people with a wide-angle of obesity, both round shoulders and sway-back are combined (kypho-lordosis).

muscles. Their increased stamina usually means that their flexor muscles are well used, but their back muscles are comparatively weak so they tend to be round shouldered, suffering from *kyphosis*.

When they are short, however, they are also likely to develop a marked arch in the small of the back in order to look up, but without a counter-balancing sacral curve. This lumbar lordosis is counter-balanced by the thoracic kyphosis, and again we have a condition of kypho-lordosis.

Narrow Angle

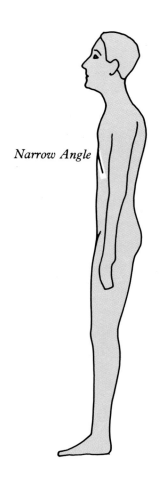

Figure 7-13. Kyphosis (round shouldered) is usually associated with a narrow angle of obesity.

Figure 7-14. In short people with a narrow angle of obesity, both round shoulders and sway back are again combined (Kypho-lordosis).

Figure 7-15. In tall people, with a narrow angle of obesity, the round shouldered slouch becomes exaggerated.

CHAPTER EIGHT

POSTURE IN CHILDHOOD AND SCHOOLS

" 'Tis Education forms the common mind,
Just as the twig is bent,
the tree's inclined."

Alexander Pope (1684-1744)

Many parents understand the need to teach their children good posture. However, the traditional method, of repeated injunctions to "sit up straight, hold your shoulders back, and don't slouch", is not very effective. Within thirty seconds the child has returned to the posture which the tone of muscles dictate.

In the child's development, it is important not to force the pace. Let him sit up when he is capable of doing so, stand when he is ready and get up and go in his own good time. The aim is not to have him walking before the kid next door, but to produce an efficient adult human being.

A major assault made on the natural good posture of children is to put them into shoes. Delay the use of shoes for as long as you can. Children are unlikely to complain of cold because their feet are well covered by fat. The skin will quickly toughen and harden, but the feet themselves will retain their great suppleness, and posture will be unaffected during early growth.

When you have to buy shoes, choose them for maximum flexibility, with a heel that can be removed easily. This is an extra expense but, particularly in children it is in fact a long term saving. Without heels the toes of the shoes are not kicked out as readily.

EDUCATION AND EVOLUTION

Going to school is a completely abnormal and illogical process. For millions of years, the school-age child spent his or her days in physical activity and play. Only during the last hundred years has compulsory schooling taken root in the west. A child of ten, who by evolutionary and biological make-up is designed to be learning the outdoor arts of hunting, fishing, agriculture and husbandry, is now often confined to a chair and shut inside a building for most of the day. It is little wonder that this enforced physical inactivity produces boredom, frustration, irritability and the beginnings of physical deterioration.

Education should instruct, train and discipline in preparation for adult life. Traditionally, this was either performed by the parents or other adults of the tribe — or by the tribe itself. Education was almost exclusively physical, interspersed with lessons in tribal cunning, customs and "know how".

The industrial revolution altered all that. Now the child is trained for a physically inactive and mentally demanding life.

Instead of learning to defend the tribe and family against animals and natural disasters, and in the last ten thousand years against other tribes, or to provide simple food and shelter — man and woman have now, to prepare to defend themselves and their family against the onslaught of technology, bureaucracy in all its complexity, noise, sight and air pollution.

Although education has changed almost entirely from the physical to the mental during the twentieth century, we cannot expect evolution to work at this speed. It would take millions of years for our body chemistry and structure to adapt to this change. It would seem logical that if our adult life, against all the principles of evolution, is to be spent in mental activity and physical inactivity, we should build up as much physical fitness as possible during school life. This backlog of fitness will help us through adult life.

It is an absolutely vital principle, (and one that has been grossly neglected, to date, in the medical field) that living material responds only to stimulus. When this stimulus is withdrawn efficiency starts to decrease. The body is economical and its efficiency is sufficient only for its needs. If physical activity is reduced the muscles will become weaker and smaller, the bones will become thinner and more fragile, and the blood vessels will become narrower and some will disappear entirely. It has been found that astronauts, in space for long periods, suffer weakening of the bone structure. This, of course, is due to the absence of the stimuli of weight bearing on the bones during their weightless flight, and is greatest in those bones which normally bear the greatest weight.

A very obvious example of the economy of the body is the speed at which milk dries up, when breast feeding is stopped. The response to stimuli is amazing! An unfit person can improve his running ability enormously in a few months, and his lifting or strength ability can be doubled. A muscle will almost double its normal store of glycogen (stored glucose) in forty-eight hours, following absolute glycogen depletion through exercise.

Education should be as concerned with physical as with mental stimuli. The young of today have an excess of mental activity and a dangerously reduced level of physical activity. School is the time to permanently affect this imbalance, by potentiating physical development. The result is a higher base level of physical fitness throughout life. Even if efforts are not made to maintain fitness throughout life, the base levels will be higher if the foundation laid down in the pre teen and teenage years are of high quality.

Bodily activity increases the efficiency of the brain by improving the supply of oxygen, nutrients and recharging it with static electricity (see Appendix — Electric Man). Academic activity does not enhance physical fitness, because it invariably takes place in conditions of physical inactivity. There is little doubt that much of the tiredness we feel after hours of mental work, is the tiredness of muscular fatigue caused by bad posture, poor desk or furniture design, and the lack of good physical conditioning.

When your child grows up he could be living in an almost exclusively chairborne society. Starting the day in a car chair, transferring to an office or factory chair, returning to the car chair and then to a television or pub chair. *Keep your child out of chairs as much as possible.* Have your television set installed in a room away from where the family normally sits, for example a bedroom. The viewer must separate from the family environment in order to watch, and only the best programmes will be tempting whilst the family living room remains just that.

We spend twenty years and many thousands of dollars educating a person for adult life, only to find that heart disease or other problems become crippling after a few years of productivity. A gold medal for inefficiency and stupidity should be awarded to such a system. We are more logical with our man-made machines. Nobody, to my knowledge, has spent twenty years working on the electrical system of a car, whilst the engine and body rotted and rusted in the rain.

EDUCATION MISUNDERSTOOD

When scholastic time has been reduced and physical time increased, it has been found that scholastic achievement has actually improved. In a 1960's French study involving several thousand children, two hours each day were changed from mental to physical education, for a randomly selected sample, and these pupils were more suc-

cessful in examinations than were those on the standard curriculum.

In an experiment in schools in Adelaide, Australia, children were given severe exercise three times a week and compared with other groups involved in less severe physical programmes. The severe exercise group produced the greatest improvement in work, confidence, health and achievement, together with an increased interest in sport and a decreased interest in television and video games.

The conclusions drawn from the experiment were that exercise must be severe to be effective and that children preferred severe exercise.

The old time British public school saw scholastic, physical, and spiritual achievement as of equal importance. Communist countries seem to be well aware of the vital role played by physical fitness in human wellbeing. The Western world is blindly going the other way. Is it that capitalism is interested in short-term profit, while communism seeks long-term gains?

The old time educator knew, and accepted, the principle of a healthy mind in a healthy body. Physical education has lost prestige through the inroads of intellectual snobbery. The physical educator is at the bottom of the education heap, and is seen as inferior by colleagues. In the free for all for a space on the school curriculum, physical education has had its time gradually eroded.

Similarly, in the education of our doctors, the scramble for curriculum time is a constant battle. The attainment of physical fitness receives not a mention. In fact, it has not even been defined, nor has its biochemical levels been measured.

Physical education is vital for posture, for without adequate muscular development and tone, good posture is unattainable. However, one of the main problems faced in physical education is to get children to like it. It is natural to like the things we are good at. The obvious solution is to help them to become good at physical activity. There is a maximum rate at which the body can improve its efficiency in response to exercise. If you increase this rate, whether in the child, the fit athlete or the old aged pensioner, then a particularly dangerous reaction occurs. Nature prefers evolution to revolution.

With a school child we must resist the temptation to push progress too rapidly, for fear of a revulsion against exercise, which might possibly become permanent. This revulsion is most likely to occur at either extremes of the angle of obesity, i.e. either in the fat, ungainly child, with poor stamina, or the scrawny weakling, both of whom most need to enjoy physical activity. Fear of physical activity prevents children from enjoying and taking part in it. For the overweight, as well as for the skinny and weak, sporting activity may be terrifying. The fat and the weak have committed no crime. Their condition is the result of heredity and upbringing over which they have had no control. By expecting them to participate, when their bodies are ill prepared, they are being punished by painful injury, or by fear of injury, or by embarrassment and derision. These fears and embarrassments colour not only their physical performance, but their academic achievement. e.g. Children are often given press-ups as an exercise. But, how on earth can fitness be improved by press-ups, if the child cannot perform even one?

There are two methods of improving strength by exercise. One is by increasing the number of times an exercise is repeated, the other is by increasing the severity of the exercise. *Weight-lifting provides the answer in a simple and scientific manner.* The weights can be tailored to the needs and abilities of the child. Increases can be made a few grams at a time, as performance improves, and the exact amount of work can be defined. In doing "free" exercises, we are working against the weight of the body. It is less controlled than if weights are used. The student is not given calculus until the mental training has been adequate, nor should he be faced with the vaulting horse, until he has had adequate training and strength development to cope.

Ridicule and derision are problems not likely to be overcome until the prowess and appearance of the child reaches average levels. If a child falls behind mentally, the cause is investigated by the School Medical Officer, the eye specialist, ear specialist, psychologist, psychiatrists, as well as the teacher, and an array of remedial help is available. If a child falls behind in physical development, little or nothing is done to help and the child falls further behind. The "play way" system is useless. The fat nine-year-old hides in the corner to do an imitation of a forward roll — the nearest and safest approach to gymnastic performance this child can attempt. Years later the child has made no progress. In academic education this would not be

permitted. With nil progress over a much shorter time, all the remedial machinery would be set in motion.

WHO DESIGNED IT?

The next assault on children's posture comes when they start school. At home, the chairs are too big and the table too high, but a normal child will not spend much time at either. At school, however, it is a different story, and it is true to say that modern school furniture design is even more harmful to children's posture than school furniture of fifty years ago.

Whilst receiving mental education the child is seated on a chair, at a desk — both of which will ruin, by their design, his good posture. The chair is too high or too low, the seat is not deep enough to support the thighs, and the back is not designed for the curves of the spine nor for variations in size. And, worst of all, the desk top is no longer slanted, but is now made flat.

It is as important for chairs to fit for size as it is for shoes, especially if the chair is to be used daily. This applies throughout life.

The height of the chair should be adjustable so that the thighs are properly supported. This occurs when the feet only just rest flat on the floor. The seat should be tilted very slightly backward, to encourage sitting right back in the chair. This helps to prevent slouching and also provides support for the thighs. The seat needs to be adjustable for different thigh lengths, so that the long legged can obtain full support for their thighs. Seats are never adjustable for thigh lengths although this is the most important adjustment of all, and simple to design. The seat can be made to slide backwards and forwards on runners, with a clamping device.

If the seat is not deep enough and is not tilted slightly backwards, much of the body weight is born by the buttocks only. With a correctly designed seat the weight is born over almost the whole length of the thigh and buttocks, so that the pressure per square inch is much reduced. Furthermore, the thighs are supported properly and do not feel uncomfortable.

Because of these design failures the spine is thrust out of line and needs back support to prevent the muscles from tiring. Modern designers try

Figure 8-1. Desk and chair too big. Nothing adjustable. Desk flat.

Figure 8-2. Desk and chair too small. Nothing adjustable. Desk flat.

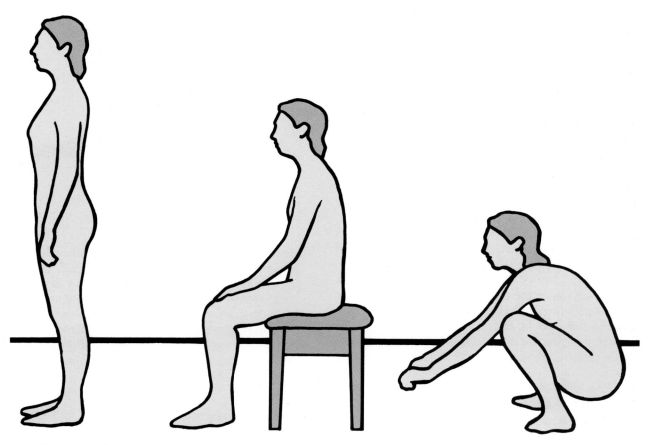

Figure 8-3. On sitting, the lumbar curve disappears and becomes slightly convex. On full squat, it becomes markedly convex.

to give support to the curve in the small of the back, but when the hip is bent as the person sits down, the curve at the small of the back disappears. We are reverting to the C-shape we had as babies, with only our cervical (neck) curve remaining. In fact, when we squat right down our lumbar curve changes from concave to convex.

If we need any vertical support, it is only for our lower backs and should be flat or C-shaped. When we sit upright, the weight above this region should be supported by our very efficient bone structure. We do not need our backs supported when we stand, and they will not need support when we sit, provided the seat is deep enough.

Desks should be adjustable for height, and the tops should be sloped, as they were years ago. It should be possible to tilt the desk top to a fairly steep angle, using modern materials to prevent papers, books, and other equipment from sliding off. The child will not have to slump forward over the desk in order to read or write.

The school child should be able to sit upright and comfortable, with the weight born by the bony frame, and not with muscles fighting a losing battle against the force of gravity.

It is muscular fatigue that produces tiredness, boredom and lack of concentration in the classroom, not mental activity. Constant shiftings of position are attempts to relieve physical fatigue and slouching is an attempt to find bony support to relieve tired muscles.

The teacher, her desk and the blackboard should be raised on a dais, to allow the pupils to see easily. Looking up helps the child to bring the body weight back over the spine. As the child spends much of his or her school life, and most of adult life looking down, the need to look up will counterbalance this tendency, (Chapter 16).

A child sitting on the left of the classroom, particularly near the front, will spend a lot of time with the head and neck twisted to the right. Children should be changed from one side of the classroom to the other at regular intervals, or classroom desks angled to face the teacher and blackboard.

It is important for small children to learn to write well, yet the tinies are expected to use the same weight of pens, pencils, brushes and other implements as adults. Infants' pencils, pens and brushes should be of infant size and weight.

If schools, classrooms, furniture and imple-

Figure 8-4. Weight born by the bony frame.

ments were designed for muscular efficiency children would learn in far less time. A concentrated mind will absorb far more than a distracted mind, racked by tired muscles and discomfort.

With postural muscles developed adequately by appropriate exercise, desks of a correct design and implements the right size, children would learn to maintain good posture without even having to think about it.

TREATMENT
See Chapter 21 for:
(A) Essential Exercises 1, 2, 3.
(B) Recommended Exercises 4-10.

CHAPTER NINE

POSTURE BULGES AND OBESITY

"I do like a little bit of butter to my bread."

A. A. Milne (1882-1956)

In the battle of the bulge, it must be terribly disheartening after weeks of dieting or jogging, or both, to find that while weight may have been lost from the face, neck, shoulders, arms, and chest, the bulges persist. Yet this is common, because the basic cause of many bulges is not overweight, but posture.

Conversely, by concentrating on exercises which improve posture, waist and buttock measurements may decrease even without weight loss. Often, when correcting posture in the treatment of such conditions as osteoarthritis of the knee or a whiplash car injury of the neck, patients have remarked that they have taken in their belt two, three or even four notches — without any conscious attempt to lose weight. There has not been weight loss, but weight redistribution.

Whether a person is underweight, overweight, or the ideal weight there remains the question of whether the weight is properly distributed. This distribution is sometimes hereditary, but normally dependent upon posture.

It is these POSTURAL BULGES which are so hard to shift, whether by diet or exercise, unless the exercises are specifically designed to correct posture. Postural bulges may truly be areas of fat, or they may be exaggerated curves without containing excessive fat. However, even in the latter, fat will eventually accumulate in these areas.

In the lordotic type, unwanted bulges will show in the buttock, abdomen, and thigh regions.

In the kyphotic, the shoulder region will be prominent, but as these people are usually thin, no obvious bulge will be apparent in the abdominal region.

In the kypho-lordotic the bulges are at the shoulder, the buttocks, the abdomen and the thighs.

All three abnormalities are exaggerated in proportion to the height of the heels normally worn.

In the regions where there are bulges, the underlying muscles are stretched, i.e. the buttock, abdominal and thigh muscles of the lordotic, the shoulder, abdominal and upper back muscles of the kyphotic, and all these muscles in the kypho-lordotic.

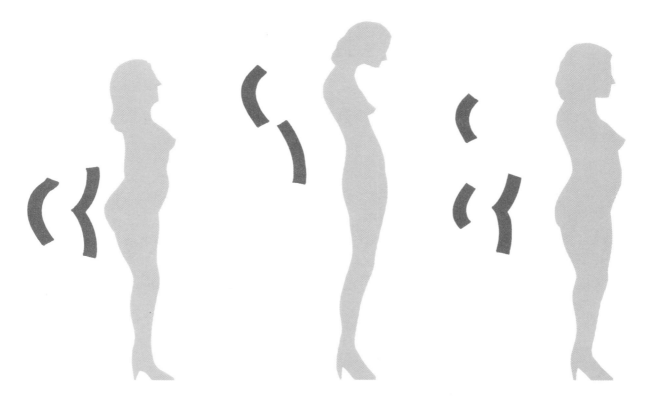

Figure 9-1. Lordotic curves.   *Figure 9-2. Kyphotic curves.*   *Figure 9-3. Kypho-lordotic curves.*

When muscles are little used, they become thinner and weaker; they degenerate. Fat starts to accumulate in and around the muscle. The build up of fat is probably influenced by the reduced blood flow which occurs when muscles are under-used.

With the passage of time, fat accumulates on the thighs, abdomen and buttocks of the lordotic.

The kypho-lordotic build up fat around the lower part of the back of the neck and their breasts become enlarged and sag. They also build up fat on the thighs, abdomen and buttocks.

In all body types, the wearing of heels restricts the range of movement of the knee and ankle joints. Because they are not used through their full range, fat and fluid tends to build up.

The most striking change occurs, however, in the tall kyphotic. Many of us remember people who were teased when they were young because they were so thin, yet, by middle age, have acquired a paunch. You can see them on the beaches in summer. Their fat bellies are quite incongruous for their arms, legs, necks, chest and buttocks remain thin, and with women because of the round shoulders the breasts will sag and tend to accumulate fat. Overeating or under exercising is not the cause. There is only one explanation – POSTURAL OBESITY.

With the passage of time, the curves of bad posture become worse, and the person becomes more bent. There is a concertina effect and height is lost. The bulges appear greater, because they are occurring over a smaller height range.

Some people may be much overweight, yet, may need to increase the weight of particular areas of their bodies. They may be bent forward in their posture, so that the bones of the spine, lacking the stimulus of compression, have become thin, light and painful (osteoporosis). They need to be thickened and strengthened, i.e. made heavier, to remove the pain. They are bent forward because the muscles which straighten the spine and the hip are comparatively weak and under developed. Most of the fat will be on the front of the body, and if the comparatively weak muscles are not strengthened first, these people will remain bent forward and the fat on the front of their bodies will take longer to shift.

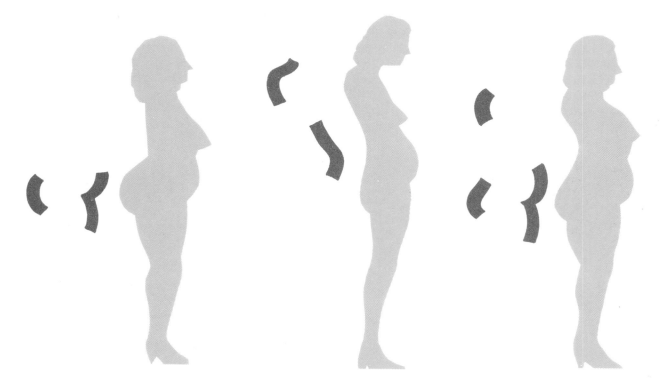

Figure 9-4. Lordotic bulges.   *Figure 9-5. Kyphotic bulges.*   *Figure 9-6. Kypho-lordotic bulges.*

> When starting on a weight loss programme, whether by diet, exercise or both, you should first correct any postural abnormalities. Only if posture is first perfected, will weight loss produce the desired result.

Before starting, the first essential is to find out whether you truly are overweight. If you look up age, height tables to work out your ideal weight, remember that with incorrect posture you will have become bent. So use the height you were aged 30 to interpret your ideal weight.

It has astounded me that most tables used to estimate ideal weight for adults have remained unquestioned for so long. These useless tables have produced mountains of unnecessary worry and concern, and erroneous estimates of life expectancy.

By reference to the age and height, the ideal weight is read from the tables.

To relate weight only to height and age is ridiculous. Width, depth and tissue density are equally important. The weight of an individual, or any other object depends upon its volume and the material of which it is made, not its length. The weight of a two metre piece of timber is not fixed. It depends whether it is a piece of twenty five cms by twenty five cms, one hundred cms by fifty cms or some other size, and the density of the timber.

It is reasonably easy to measure height accurately, but the measurement *must* be taken in bare feet, for heels will buckle the posture, and distort the true height reading. An estimate must also be made of the loss of height due to bad posture.

To measure the width and the depth of the body is a difficult problem. Although there are people with big chests and narrow hips and legs, and vice versa, — measurement of the circumference of the chest will give a reasonably accurate measurement of the size of the bony framework. The effect of posture on this measurement is minimal, for there is a tendency to lose on the swings what is gained on the roundabouts. A better method to calculate width and depth would be to measure the angle of obesity (Chapter 7). This gives a ratio between height and width, and is related to the size of the bony framework. While this angle takes no direct account of the depth of the frame, deep chests normally go with wide chests.

The problem of tissue density is just as complex and requires amongst other things the measurement of body fat and musculature.

Until logical tables are statistically compiled, masses of people, believing incorrectly that they are overweight, will fight a hopeless battle trying to maintain a weight lower than their optimum, — with worry, guilt and unnecessary misery.

The mirror used critically and honestly when naked, will give a simple and fairly accurate answer as to whether there is general obesity — or whether there is local (i.e. postural) obesity — or both.

The medical profession expresses little con-

Figure 9-7. "You are in perfect condition, — but, unfortunately, you are 16 kg. overweight!"

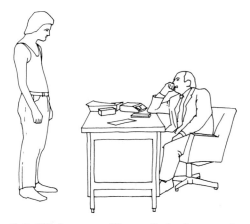

Figure 9-8. "Take your pills regularly for your bronchitis, but you are 9 kg. underweight, so you will have to eat more!"

cern for those who are markedly under the weight that tables state they should be. No amount of gorging or exercise brings these narrow chested people up to their predicted weight, unless they develop postural obesity, when they will be deluded that their fat abdomen is of no significance, because they have reached the so called "ideal weight".

The number of authorities on obesity and diet probably exceed the number who are obese. They have managed to turn a simple subject into an extremely complicated one. When the calorie intake (food) exceeds the calorie output (energy used) weight is gained (fat storage) at a rate dependent upon individual characteristics and that is all there is to that!!

Some people absorb food from the gut more easily than others, either because the gut lining is more efficient, or the gut is longer and bigger. When the food has been absorbed, there are differences in efficiencies with which individuals utilise it. Nevertheless, the basic principle applies that weight is a balance between energy eaten and energy used up.

> **Exercise is the great leveller. It reduces the weight of the obese, and increases the weight of the skinny.**

Without exercise, the obese will continue to get fatter, for they are caught in a vicious cycle.

When they put on weight through lack of exercise, they tend to exert themselves even less, because of the excess weight they have to carry. This allows the weight to increase even further. Those with an extremely wide angle of obesity are likely to end up with a way of life, where exertion is reduced to performing the bare necessities.

Invariably, the thin have stamina and are energetic. Their only way of gaining weight is to increase muscle size and bone density. This can only come from exercise.

As weight can only be lost by reducing the energy input (food), or increasing the energy output (exertion), or both, we can make a choice. Of the three increasing the output with physical activity is usually the more pleasant and successful.

On its own, a low calorie diet will normally produce only temporary weight loss. The diet is always started with great enthusiasm and determination. Two or three kilograms are often lost in the first week, much of which may be due to fluid loss. The next week the loss is usually a lot less, and the diet begins to pall. The first dietary indiscretions occur as will-power fails. By the third week, the loss may be minimal. The battle to fight temptation becomes harder, and the loss by the fourth week negligible. Eventually the diet is abandoned, because there seems little point in continuing the agony, without losing significant weight. Having rejected the diet, the person often gorges and returns to the original weight or higher. This tale of woe is shown in the Failing Graph of Diet.

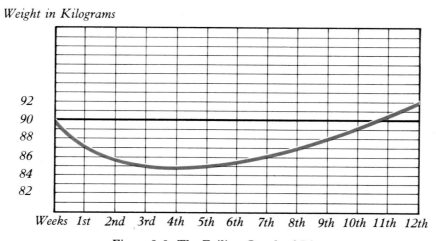

Figure 9-9. The Failing Graph of Diet.

Exercise may be a slower way to control weight, but, like the tortoise, it is a winner in the end, and has the added bonus of producing increased fitness and improved health. The biggest problem in prescribing exercise for weight reduction is to convince the victim that it will work. People have lost faith in the use of exercise, because of some fallacious experiments. It has been stated that so much walking, or so much running, is equivalent to a certain quantity of food. The fallacy of this argument is that the measurements are obtained only during the exercise period. The benefits, however, in terms of energy consumption continue for several days after the completion of the exercise. Muscles ache for days after severe and unaccustomed exercise, and until those muscles have fully recuperated, they are using up energy. Even the fit athlete uses energy to maintain muscle tone for hours after a training session.

Before significant weight can be lost with exercise, one must attain the fitness to be able to do sufficient exercise. Results are not immediate, and, without overtraining, there is only a certain rate of progress which can be made, which varies for each individual. If attempts are made to progress too rapidly, injuries or even heart trouble may ensue, and a revulsion against further exercise is highly likely. The obese must commence very gently, and remember the excess weight that they are carrying. Fit people could not exercise for long if they had 25 extra kilograms strapped to their back. As fitness improves, the amount of exercise increases and so does the weight loss. This is shown in the Succeeding Graph of Exercise.

A further bonus is that, as fat is lost, muscle tone improves and the skin and underlying tissues become firm. There is less unsightly sagging of the kind that is seen in people who have been excessively overweight and have relied solely upon diet. Skipping is a cheap, convenient method and can often be done secretly at home — avoiding any embarrassment. For women, especially, it is a good way to start exercising — but it soon becomes boring.

It is fashionable now to advise jogging. But, obese people have very little stamina, and often loathe jogging and are unlikely to persevere with it. The naturally obese are more suited to some form of interval training to improve cardio-vascular efficiency and to lose weight, than to continuous stamina activity, such as jogging. If these people were to run in short bursts, with walking in between, they could cover adequate distances without the severe discomfort that continuous activity brings for them.

Although they have poor stamina, they have great strength and like to use it. They are ideally suited to weight-training, which they often take to with relish and success. There is, unfortunately, still some aversion to this form of exercise. There is a fear, especially amongst women that they will become muscular and unfeminine. This is a myth, which is being gradually exploded. For the big woman, weight-training is, by far, the best method of improving her figure.

There seems to be little place for surgery in the

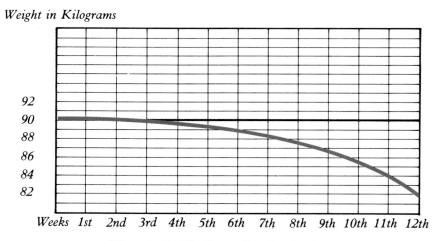

Weight in Kilograms

Weeks 1st 2nd 3rd 4th 5th 6th 7th 8th 9th 10th 11th 12th

Figure 9-10. The Succeeding Graph of Exercise.

treatment of the obese. The operations carry a considerable risk, with long term complications a possibility — a major assault upon the body for something that can be so successfully treated with exercise.

It is worthwhile for the obese to wear fewer clothes. In winter, if you are underclothed, your body will burn up more food to maintain its temperature. And if you feel cold, you are likely to become more active in order to warm up. One old sage, during a power strike, remarked that he did not worry if the central heating was not working; he had the best central heating system in the world, — right inside himself.

> **Whether obese; appearing obese because of postural bulges; or obese in parts of the body only, due to postural obesity; correction of posture is TOTALLY IMPORTANT.**

If bad posture is allowed to remain for years, sooner or later osteoarthritis in a weight bearing joint (Chapter 12); back pain (Chapter 13) or a posture induced injury will occur. When this happens, activity will be greatly reduced and obesity will be accelerated. With a supreme effort, most of us can tuck our tummies in. The problem is to hold the tummy there. This becomes effortless, when posture has been corrected with appropriate exercise.

TREATMENT
See Chapter 21 for:
(A) **Essential Exercises 1, 2, 3**
(B) **Recommended Exercises 5-10**
(C) **The round shouldered kyphotic should add Exercise 4**
(D) **The sway backed lordotic should add Exercise 22, 23**

CHAPTER TEN
POSTURE IN MATERNITY

*"It is as natural to die as to be born;
and to a little infant, perhaps the one
is as painful as the other."*

Sir Francis Bacon(1561-1626)

The pregnant woman is likely to think that she has enough problems to worry her, without being concerned about how she stands. However, if she were aware that this stance was going to play an important, possibly vital part in the ease and safety of delivery of an undamaged baby — and that she, herself, could solve the problem easily — she would certainly consider it.

The physical act of having a baby is a simple, mechanical procedure, but the ever-increasing weight in the tummy as the child grows upsets normal balance and affects the channel through which the child passes during delivery.

Good posture is essential if she is to revert to her normal body size and shape, after delivery. This is important not only for her own self satisfaction, but also to help in any future pregnancies.

Many children are born with handicaps that could have been prevented. Even in countries with the best maternity care, babies are born with brain damage caused by poor oxygen supply to the brain or by haemorrage inside the skull from forcep deliveries. I believe that many of these children could have been saved from handicap by the use of simple exercises to adjust the posture of the mothers during pregnancy. A minute price to pay to avoid the tragedy of brain damage, and to ensure the safe arrival of that most fascinating, most intriguing, most amazing, most lovable creature — a perfect baby.

THE MECHANICS OF NORMAL DELIVERY

The major problem is to get the baby's head through the mother's pelvis. When labour begins, the baby's head sits over the brim of the pelvis, with the head lying in the side to side (lateral) position.

The forces produced by the muscular contraction of the uterus (womb) act on the baby causing its head to bend forward. This changed angle allows a smaller diameter of the head to pass through the pelvis.

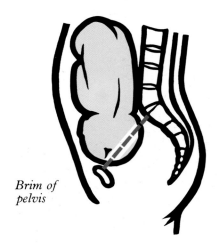

Brim of pelvis

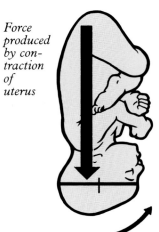

Force produced by contraction of uterus

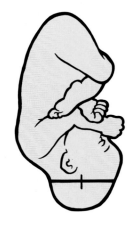

Figure 10-1. Usual position of baby at commencement of labour.

Figure 10-2. Diameter of head to pass through pelvis, before bending of neck.

Figure 10-3. Diameter of head to pass through pelvis, after neck has bent (flexed).

As the head descends through the pelvis it rotates with the result that the back of the head comes to the front of the pelvis.

Further pressure causes the head to extend on the neck (the head bends backwards), and the face and chin pass through the outlet of the pelvis.

As the head passes through this bony outlet it stretches all the soft tissues — the anus, the vagina and the tissues between them, called the perineum. Eventually, the head passes through the vagina. The shoulders follow the same sequence of rotation as they pass through the pelvis and normally, no great difficulty is encountered in the delivery of this, or subsequent parts of the baby.

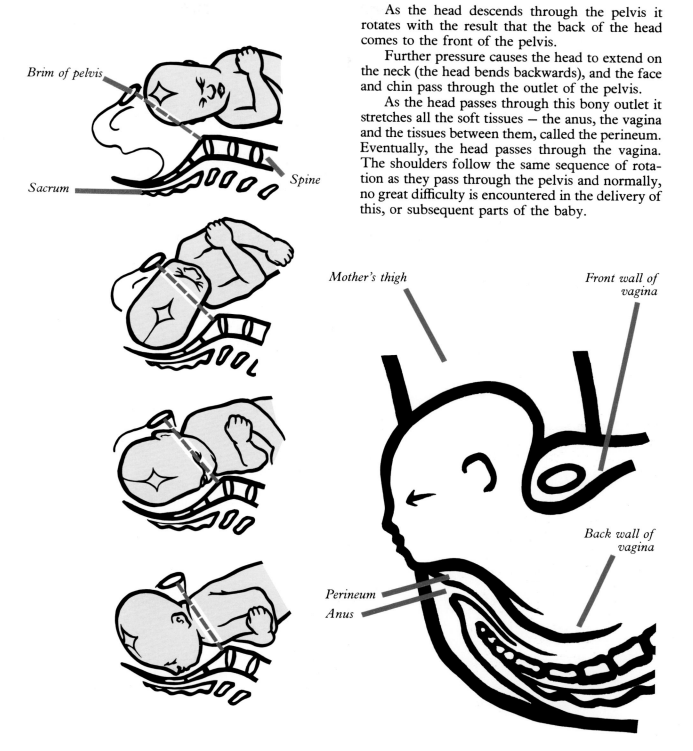

Figure 10-4. Head descending and rotating as it passes through pelvis.

Figure 10-5. Head, face and chin are delivered.

The uterus, in which the baby has developed, is a bag with thick, muscular walls that has a special lining to succour the developing egg, and a profuse blood supply. At full term, the uterus has grown enormously, and the muscular wall has become much thinner. The baby is suspended in the water (amniotic fluid) which is retained inside the bag (the membranes).

included as an occasional exception.

Posture has been largely ignored in the field of pregnancy and delivery. In a list of the factors affecting successful labour, it normally warrants no mention. Yet its importance is vital.

Obstetricians and gynaecologists, have their eyes focused on one region — the pelvis. They do not see the total body, although not all are as bad

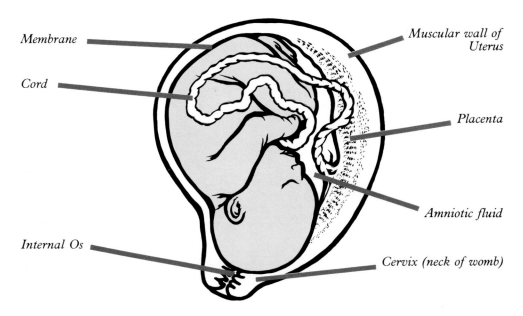

Figure 10-6. The uterus (womb) at full term.

Membrane

Cord

Internal Os

Muscular wall of Uterus

Placenta

Amniotic fluid

Cervix (neck of womb)

Once labour starts, the pressure of the contractions causes the cervix (mouth of womb) to open up. Eventually, when the second stage begins, the cervix is fully opened and the mother has an overwhelming desire to bear down on the baby and to expel it from the uterus.

The uterus works with a squeezing action, but, when it relaxes between contractions, it does not return fully to its previous size, so that the baby does not slip completely back each time the muscle of the uterus relaxes. Delivery should be a steadily progressing procedure.

POSTURE AND OBSTETRICIANS

It is remarkable, that the complete woman is entirely absent from most obstetric textbooks. Photographs, diagrams, figures and x-rays are usually exclusively of the pelvis, with the abdomen

as one famous gynaecologist, with a notoriously bad memory. He was unable to recall any of his returning patients until he had examined the cervix, when he was wont to say "Oh, I remember this case!"

THE EFFECT OF PREGNANCY ON POSTURE

The weight of the baby, the enlarged uterus, and the amniotic fluid upset the woman's balance and she is forced to lean backwards in order to keep her centre of gravity over her feet. The round-shouldered woman tends to achieve this by thrusting her hips forward and rounding her shoulders even further. The hollow backed woman, on the other hand, solves the problem by increasing the arch in her back and sticking her bottom even further out.

If, in addition we put both women in shoes with heels, they will have to make adjustments for a forward tilt which will range from twelve to fifty degrees, some of which will be accommodated by the ankle joint.

The woman with the arched back will correct the remaining tilt by bending at the knees and, to a lesser extent, at the hips. She will also increase

common in whites as in blacks; while the opposite, in which the front-to-back diameter is lengthened, is twice as common in blacks as it is in whites. It is possible that the variation in the shape of the pelvis is due to the wearing of shoes with heels rather than to racial or genetic differences.

The tall, round-shouldered (kyphotic) type of women, and those with good normal posture,

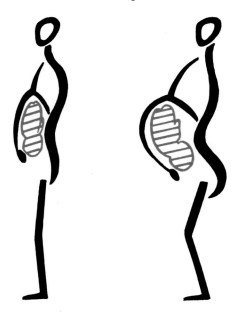

Figure 10-7. The round shoulder (kyphotic) (left) and the arched back (lordotic) (right) counterbalancing the weight of the baby.

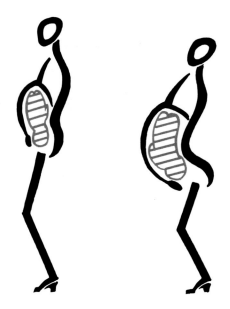

Figure 10-8. Pregnant kyphotic (left) and pregnant lordotic (right) in shoes with heels.

the curves of her spine in the small of her back and in her neck, so that her bottom and her belly will stick out even further. And it is the short woman with the increased curve in the small of her back, who is most likely to wear raised heels.

The round-shouldered woman will also exaggerate her curves similarly increasing the distortions in her weight bearing joints. However, she is luckier in that the position of her child will remain helpful to delivery.

POSTURE AND DELIVERY

The excessive curves often cause the spine to buckle at the lower back, producing a narrowing of the entrance to the pelvis, by shortening the front-to-back diameter. This type of pelvis is twice as

should carry their babies well, and their girth will not be excessive.

The short, hollow-backed (lordotic) type of women are likely to develop large, pendulous abdomens, with the muscular wall of the abdomen excessively stretched. This stretching becomes greater with each succeeding pregnancy.

It is in these women with large, pendulous abdomens that malpresentations tend to occur, i.e. when the face, shoulder or buttocks (breech birth) attempt to pass through the pelvis first. Malpresentations are amongst the most difficult and dangerous of the complications of childbirth. They are three times more common in the fifth than the first pregnancy. One cannot explain this difference by claiming a narrowed pelvis. The answer is in deteriorating posture. Malpresentation is a complication that could be largely prevented.

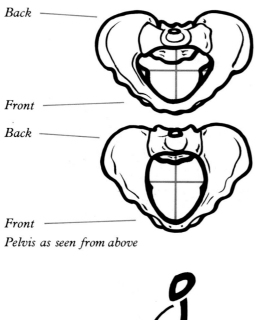

Back

Front

Back

Front

Pelvis as seen from above

Figure 10-11. Pendulous abdomen in sway-backed (lordotic) woman.

Figure 10-12. Increasing stretching of abdominal muscles with successvive pregnancies in lordotic women.

Figure 10-9. Pelvis seen from above. Shortened front to back diameter, i.e. Android pelvis.

Figure 10-10 Pelvis seen from above. Lengthened front to back diameter, i.e. Anthropoid pelvis.

It is usually found that excessively big babies are born to mothers who are obese and have had two or more pregnancies. The arched back types with their large pendulous uteri have the room to accommodate such babies. Statistics show that these large babies are at greater risk during the first few days of life than normal sized babies.

Women who are thirty-five years or older and who are having their first babies, more frequently have longer labour, and babies born to these women have a mortality rate that is three times greater than for babies who are born to younger women. The explanation that is usually given is that the muscle of the uterus does not function properly and that the cervix is apparently more rigid. However, posture has often become worse with ageing and it is probable that it is this bad posture which has reduced the efficiency of the uterus, and that this in turn, slows the opening of the cervix.

During pregnancy, hormones make the ligaments of the joints more pliable, so there is more flexibility of the joints, especially the pelvis. This adds further to the problems of posture, for it makes distortion more likely.

As the bad posture of the hollow, arched back woman becomes worse during pregnancy, the muscles in her large pendulous abdomen may become severely stretched and damaged. This, in turn, may lead to the foetus lying in the wrong position in the womb, and the further complication of the umbilical cord presenting first because the head is not fitting snugly at the opening of the pelvis.

Problems can also arise during labour as a result of poor posture. With each contraction the uterus becomes straightened and lengthened so that the abdomen is pushed forward. The expelling force of the uterus acts down the length of the foetus.

In a woman with good posture, or of the round shouldered type, the curvature of the lower spine is gentle, so that the curves which the foetus follows through the pelvis are also gentle, and the expelling force of the uterus acts in a beneficial direction.

Where there is excessive curvature in the lower back, the path of the foetus through the pelvis is markedly curved.

Figure 10-13. With bad posture, the umbilical cord may be below the head during delivery — a dangerous complication.

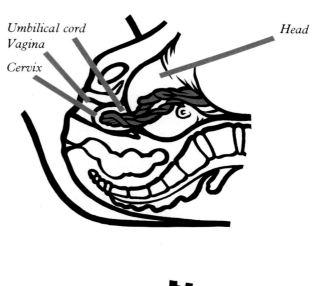

Umbilical cord
Vagina
Cervix
Head

The baby's head tends to hit against the tail bone, and a large part of the force of the contraction is wasted. Only a relatively small part of the force is pushing the head through the pelvis. Most of the energy is being wasted in ramming the head against the tail bone. This impact against the tail bone cannot do the baby any good, and the contractions may last for many hours, even days. This pressure of the head against an unyielding surface helps to produce swelling, bruising and distortion for there is an equal and opposite force acting on the babies head. If the force of the uterine contraction is resolved into the useful force propelling the baby through the pelvis, and the wasteful force compressing the baby's head against the sacrum, it is seen that, if the curves are shallow, the useful, expelling force is maximum.

In the hollow backed woman, the damaging force is not only much greater, but acts on the baby's head for a longer time, since these women have longer labours.

We have yet another complication in labour for these women. The space available for the rotation of the babies head during its passage through the pelvis is much reduced sometimes resulting in incomplete rotation.

Difficult labour is considered to be the result of bad functioning of the muscle of the uterus, abnormalities in the size and shape of the uterus or abnormality of the foetus. Except for damage by previous operations or deliveries, the uterine muscle is unlikely to be abnormal, and its function is also unlikely to be impaired. It is the position of the uterus which is the usual culprit, not weak contractions. If this useful force is sufficiently reduced, the muscle of the uterus may become completely exhausted before the baby is delivered, and all contractions may stop. Muscles stop working when their store of energy (glycogen) is drained, and it is wise to take some form of sugar during labour, so that the muscle of the uterus can be continuously re-energised.

Arched back posture is the most likely to pro-

Figure 10-14. Direction of force of action, when the womb contracts.

Figure 10-15. In the normal and kyphotic postures expelling force acts in a favourable direction.

Figure 10-16. In the lordotic the expelling force acts in a less favourable direction, and the baby has to make a more tortuous path through the pelvis.

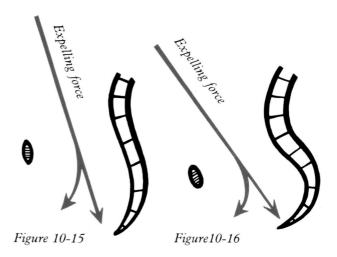

Figure 10-15   *Figure 10-16*

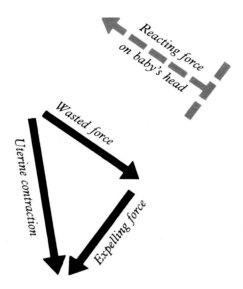

Figure 10-17. Forces acting on baby's head in kyphotic. Maximum expelling force.

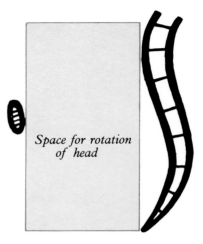

Figure 10-19. Space available for rotation of baby's head in kyphotic.

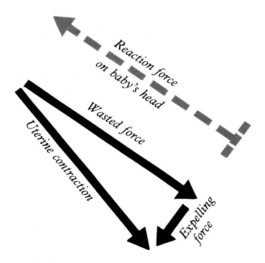

Figure 10-18. Forces acting on baby's head in lordotic. Minimum expelling force. Maximum damaging force on baby's head.

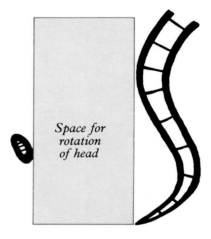

Figure 10-20. Reduced space available for rotation of baby's head in lordotic.

duce difficulties in the size and shape of the pelvis. Congenital factors, previous injuries to the pelvis, and past diseases such as polio or rickets may all distort the pelvis, but these are usually assessed and allowed for, well before delivery, and do not form part of a discussion on a normal delivery.

If there is an excessive curve in the sacrum the exit from the pelvis is reduced in size, and this is likely to occur in the arched back type.

Even if the baby is born unaided in these women, there is likely to be gross stretching, tearing and distortion of the anus, perineum, front and

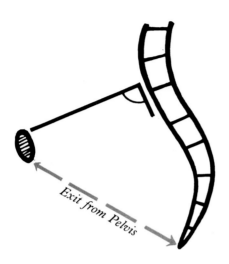

Figure 10-21. Wide exit from pelvis (pelvic outlet) in kyphotic.

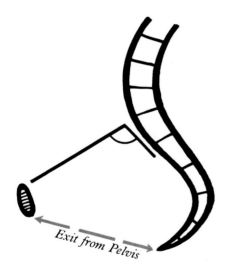

Figure 10-22. Narrow exit from pelvis (pelvic outlet) in lordotic.

back walls of the vagina and the urethra, and these injuries may require immediate repair, and may even produce permanent disability.

Evidence has been found to show that African women have easier deliveries, with less complications, than the women of industrialised coun-

tries. The African mother has many advantages, mainly postural, that arise because she does more physical activity than her industrialised counterpart and if she wears shoes, they are likely to have no heels or very low heels.

It has been suggested that athletic women have difficult deliveries because their powerful muscles cannot relax. This is a false assumption! The ability of a muscle to relax is not related to its power. In fact, top competitive weight-lifters have great powers of muscular relaxation and flexibility. The arched back posture however can arise in a sportswoman because the muscles of the lower back are out of balance with other postural muscles.

I would not be apprehensive about attending the deliveries of a team of international basketball or volley-ball players who are likely to be tall and slim, with their posture improved by repeated reaching and jumping.

TRAINING FOR CHILDBIRTH

First *get out of shoes with heels.* I know of only one condition for which doctors recommend shoes with low heels, and that is pregnancy, so there has been some recognition of the problem. Even one centimetre heels will thrust you forward about eight degrees. This distortion of posture may make all the difference in the ease of your delivery, so remove even that one centimetre.

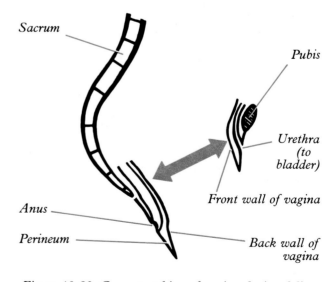

Figure 10-23. Gross stretching of vagina during delivery.

Any pregnant woman will benefit from posture-correcting exercises, but they apply especially to those with arched backs. If you are short, you are more likely to be of this type, but you can test yourself. Take your shoes off and stand with your back against the wall, and your heels about ten centimetres out from the wall. Bend forward so only your bottom is touching the wall.

Come back slowly against the wall, trying to press against it with the entire length of your spine as you come up, until your head touches.

The only gap between the wall and your spine should be in the region of your neck. If there is a gap in the small of your back — the lumbar region — then postural exercises will be especially helpful.

Treatment with exercise has one distinct advantage over treatment with drugs; the doctor can tell easily when it has been taken. If exercise treatment is carried out, there is improved performance and with experience, the degree of improvement can be forecast and measured.

For ten years in a busy practice, I used exercise to improve the posture of my pregnant patients. Out of hundreds of deliveries, one was delivered by caesarian section because of an enlarged head from hydrocephalus (water on the brain); one was delivered by forceps, because the mother had a lobe of her lung removed and she rapidly became distressed during delivery; and a third was delivered by forceps because of delay in the second stage, caused, I believe, by an arched back posture. This mother had her first ante natal examination shortly before she was due to deliver and therefore, no exercises. The remainder had straightforward deliveries without complications and without forceps.

> **In maternity, strong abdominal muscles to maintain the uterus in its correct position are essential.**

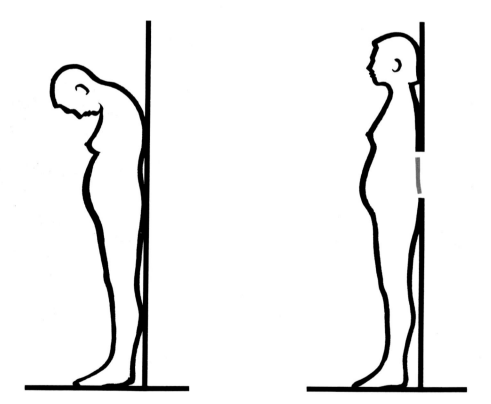

Figure 10-24. Testing for good posture. With the heels 10 centimetres from the wall, it should be possible to flatten the lumbar region.

I delivered most of the women on their sides. The delivery position with the woman lying on her back with her legs raised, has some postural advantage in that it tends to flatten out the back. However, it has disadvantages. It tends to be uncomfortable and tiring, and many women object to it because it further decreases their active participation in the birth and the feeling of control over themselves, that is so important. The most logical delivery attitude would be the squat position, and this is sometimes used by women when they are given the choice. However, it is a difficult and tiring position for many women and awkward for the person assisting.

The squat position can, however, be adapted to the horizontal by using an adjustable, padded couch that has a continuous C-shaped curve. There should be foot rests on which the woman can press during contractions, and hand grips to further assist her bearing down. Shoulder pads would ensure that she did not straighten her knees.

The force of gravity plays no part in a delivery, so there is no advantage, other than postural, for the mother to be in the vertical position.

Generalised exercise to improve the cardio-vascular system, such as running, swimming, or cycling, should be encouraged. However, the only specific exercises which will help and not hinder delivery are of an abdominal and postural nature.

Women are at last, recognising the benefits of exercising with weights as opposed to free exercises which use the weight of the participant's body as resistance. Women who weight train can continue their training providing they avoid exercises which flex the hip or arch the back.

TREATMENT: See Chapter 21 For
(A) **Essential Exercises. 22, 23.**
(B) **Recommended Exercises. 1, 2, 3.**
(C) **Helpful Exercises 5-10.**

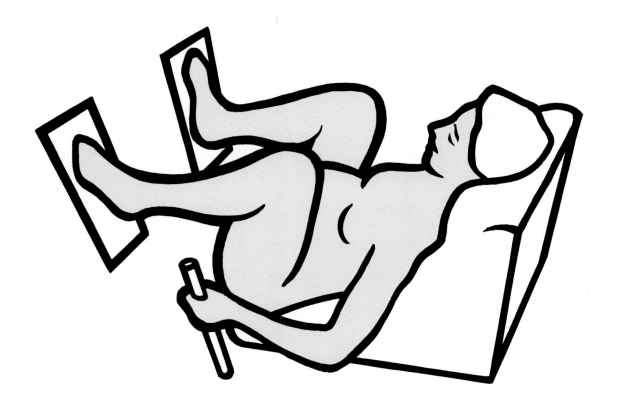

Figure 10-25. The squat position, posturally ideal for delivery, adapted for the horizontal.

CHAPTER ELEVEN
POSTURE AND APPEARANCE

*"Vitality and beauty are gifts of nature
for those who live according to its laws".*

— Leonardo da Vinci (1452-1519)

PHYSICAL

Tell a horse by its teeth, a tree by its rings, and if you want to judge the age of most humans, look at their necks.

The appearance of the face can be altered by the amount of fat. It can be creased by worry and depression, smiling and laughter. Sun or close work can screw up the eyes; sport or exertion can clench the teeth and thin the lips. The face is not a reliable gauge of age; neither is the hair, which may be coloured or grey, thick or thin at any age. Hands are unreliable, as their skin texture varies with the type of work they do. Feet are distorted by unsuitable footwear, lack of physical activity, the angle of the knee joint, and poor circulation. The position of the head on the neck however, is largely dependant upon posture and because poor posture becomes worse with age, the appearance of the neck has a direct relationship to the age and posture of the owner. Few abnormalities affect this relationship. Distorted teeth may alter the shape and position of the lower jaw, varying the skin tension of the neck or goitre (swelling of the thyroid gland) may change the shape of the neck, but generally the neck is a reliable age indicator.

In both types of postural abnormality — kyphotic (round shouldered) or lordotic (sway or arch backed) — the normal curvature of the spine in the region of the neck is altered.

With both types, there is an initial stretching of the neck. With age, this causes sagging lifeless skin with a parchment like appearance, and the back of the neck becomes creased.

The solution to sagging skin is not the plastic surgeon's knife, but lies in early correction through postural exercises and correct footwear.

The double (or treble) chin is largely a postural condition. It is common in both postural types, but especially in the lordotic types where the neck is usually held back excessively.

This position tends with age to become rigid, so that when they look forwards or downwards, they bend their heads by tilting the head on the top of the spine, with very little movement in the neck itself. The skin underneath the chin folds, fat accumulates and the double chin is produced.

The typically tall kyphotic type has an increased cervical curve to counterbalance an increased thoracic curve. (Their slouched posture would make the head look straight down if the neck were not tilted backwards). The neck again tends to become rigid, so that when they wish to look up, they nod their head backwards using mainly the joint between the head and the first bone in the neck (atlanto — occipital joint). Their jaws tend to become thrust forward and jutting, stretching the skin of the front of the neck and creasing the skin at the back. The cartilege surrounding the voice box is pushed forward producing a prominent Adam's Apple common in these people as they age.

The only part of our faces affected directly by posture is the area around the eyes. Our modern way of life, reading, writing, typing, at the work bench — necessitates looking down most of the day causing the lower eye lid to become creased and baggy.

The height of a tall person may produce a slight stoop. Age increases the stoop, and fat may be deposited in the lower part of the back of the neck and over the shoulders. The extreme form of this has acquired the name, Dowager's Hump. Although plastic surgeons tackle many cosmetic

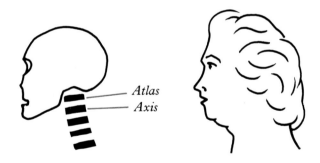

Figure 11-1. Age sees the development of the double chin in the lordotic. (arched back).

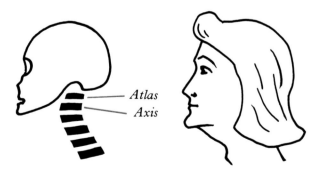

Figure 11-2. The jutting jaw and protuberant Adam's Apple in the kyphotic (round-shouldered).

defects produced by poor posture, they have little to offer for this disfigurement. Postural therapy will generally correct the problem painlessly and cheaply.

If excessive curves are allowed to remain, osteoarthritis of the neck (cervical) and chest (thoracic) parts of the spine are likely to occur. The bones of the spine tend to collapse and the lungs will be cramped for space with restricted ability to expand.

In women who have the narrowed, depressed chest of the round-shouldered, the breasts sag and become pendulous. The breasts are often small in the young, but round shoulders restrict localised muscular activity in this region, circulation is decreased, fat is deposited and they may become greatly enlarged. Although good posture will prevent this happening, correcting the posture will not completely cure an established condition.

The round-shouldered seldom suffer the problems of a fat bottom. Theirs are usually small and unobtrusive. However, with increasing age the skin and fat over the almost non-existent gluteal muscles of the buttocks sags unattractively. A full, rounded bottom is one of natures great charms, and can be obtained and maintained by postural therapy.

The ravages of age and poor posture are much greater in the sway-backed (lordotic) type of person. There is, however, a marked sexual variation caused by the differences in the femur (thighbone). In the male, the neck of the femur is not as horizontal as in the female and the head of the femur is inserted in the hip joint so that the neck of the femur is angled backward from the joint. In the female, the neck of the femur comes out horizontally, sideways. This means that the hips are wider in the female, and is the reason why many women

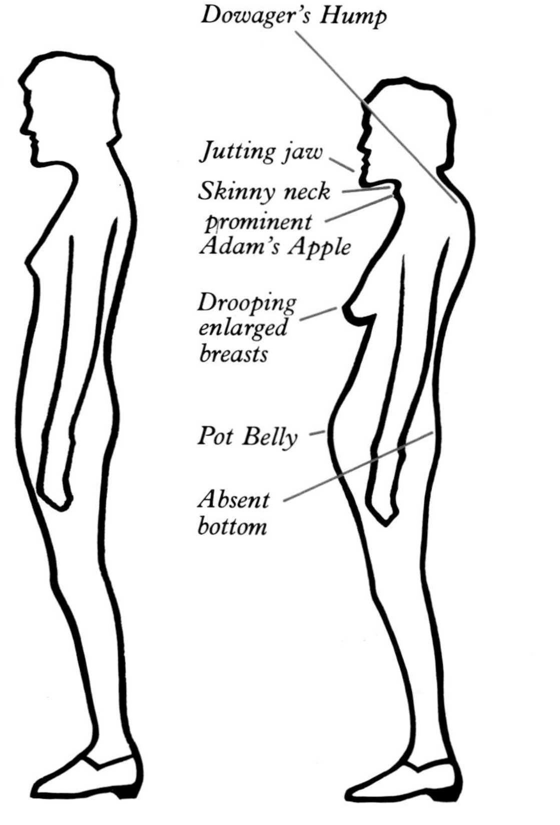

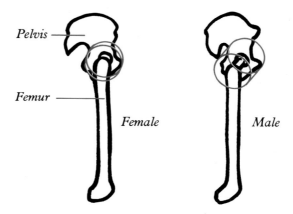

Figure 11-4. Thigh bone (femur) and pelvis seen from the right hand side. The neck of the femur is angled backwards from the joint in the male, making his hips narrower.

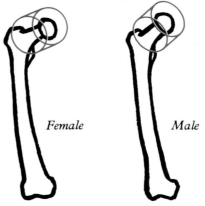

Figure 11-5. Thigh bone (femur) seen from in front. The neck of the femur is nearer to the vertical in the male, making him taller, and, again, making his hips narrower.

Figure 11-3. Changes developing in kyphotic with age.

Dowager's Hump

Jutting jaw

Skinny neck prominent Adam's Apple

Drooping enlarged breasts

Pot Belly

Absent bottom

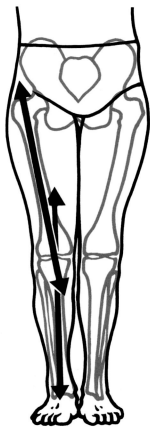

Figure 11-6. The wide hips of the female often make women appear knock-kneed, when they are not.

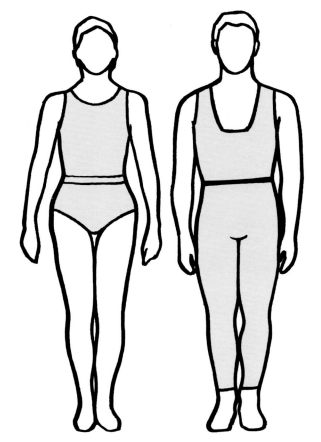

Figure 11-7. Triangular gap at the top of the inner side of the thighs in the female and none at the top of the male.

appear to be knock-kneed (when in fact they are not) and why they have a small triangular space at the top of the inner side of the thighs.

This evolutionary sexually oriented modification of the femur in adapting from four-legged to two-legged posture had advantages for both male and female. It pushed the males testicles forwards, stopping them from being squeezed between his legs. It made him taller, to see further when hunting. It increased his stride length, so that he could run faster and jump further, and it enabled him to move more easily, lower and faster, when hunting in a crouched position. The female femur made squatting easier for her, in her daily chores around the tribal home, and in the delivery of her baby. These differences were and are vital to both partners during sexual intercourse (see Chapter 15).

With the female thigh bone set forward, the curve in the small of the back is greater in order to bring the chest further back to counteract the forward weight of the thighs. Furthermore, the shorter the person the greater the curve in the small of the back. (Chapter 5). With advancing years, the dainty arched back physique of the young deteriorates. Gravity produces an increasing muscular imbalance and this in turn if unchecked by postural therapy exaggerates the arch in the lower back. The buttocks project further and the little used buttock muscles become enveloped in thick layers of fat; the abdomen, breasts, and chin sag and accumulate their share of excess fat; a spare tyre develops; the muscles on the front of the thigh never contract fully to straighten the knee, and so the thighs join in the obesity; fat accumulates behind the knees. Because the knees are continually

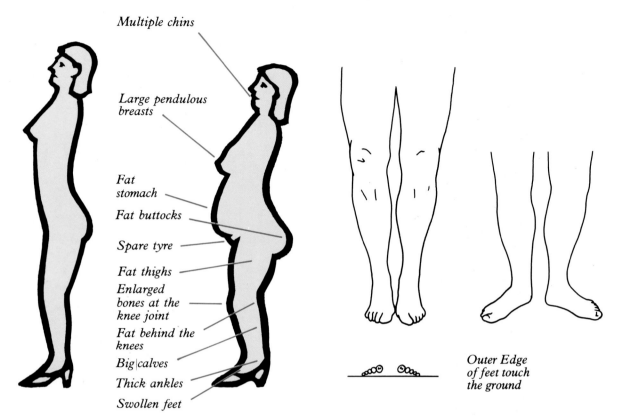

Figure 11-8. Changes developing in lordotic with age.

Multiple chins

Large pendulous breasts

Fat stomach

Fat buttocks

Spare tyre

Fat thighs

Enlarged bones at the knee joint

Fat behind the knees

Big calves

Thick ankles

Swollen feet

Outer Edge of feet touch the ground

Figure 11-9. Bandy person (a) before making any adjustments of the foot, ankle and knees for bandiness and (b) after flattening the feet, turning them out, and bending at the knees.

bent, the bones of the knees *appear* bigger. If they develop osteoarthritis which they are likely to do, then the *actual* size of the knees will increase.

Shorter than average women tend to wear higher heels, shortening and enlarging their calf muscles and thickening their ankles. One of the major activities of the calf muscle is to squeeze the veins inside as it contracts, forcing the blood in the direction of the heart. With the reduced movement of the calf muscle due to high heels, circulation becomes less efficient and fluid collects in the tissues in the same way as it does when we sit for long periods in an aircraft or on any long journey. The short person with a sway back is thus even more likely to develop swelling of the ankles and feet, and if this occurs over a long period, permanent swelling of tissue may result.

If a bandy person makes no postural adjust-

ment to counteract the bandiness, all the weight is supported on the outer edges of the feet. When one leg is lifted, in running or walking, the weight is born on a very narrow outer strip of the other foot, and the person tends to topple outwards. To keep balance on one foot, the bandy person flattens the foot to bring more of it into contact with the ground, points the foot outwards and bends the knee forwards and inwards, to bring the foot under the centre of gravity, a stance typified by Charlie Chaplin. The bandy person is inclined to be flat footed to counteract the bandiness. Arch supports, shoe inserts, and building up the sides of the shoes serve only to mask and perpetuate the original problems of bandiness.

Most first-class sportsmen and women have a tendency to bandiness, brought about by having to frequently bring the leg toward or across the midline, using the inner leg muscles (adductors). This is particularly noticeable amongst soccer players.

The correction of bandiness by exercise would require considerable and possibly unjustified effort. But, as bandiness will become worse with age, and increase the likelihood of osteoarthritis of the knees, preventive exercise is worthwhile. Slight bandiness is of benefit to the sports person, for the knees never get in the way, and the leg can be brought straight forward without knocking the other knee.

The problems of the knock-kneed are generally the reverse of the bandy. Without an adjustment at the ankle or the foot, the weight is borne down the inner side of the foot and the big toe, making the person liable to fall inwards when standing on one leg. So the foot is inverted to bring the whole weight-bearing surface of the foot into contact with the ground, and the foot is turned inwards (pigeon toed) bringing it under the centre of gravity to avoid toppling inwards when the person is standing on one leg.

At first sight, the bandy person may appear knock-kneed because of the forward and inward bend of the knees, and the knock-kneed person to appear bandy because a pigeon-toed appearance produces a gap between the knees. To check whether or not your legs are straight, try to put your ankles together while keeping your knees fully locked backwards. If you are bandy, there will be a gap between your knees. If you are knock-kneed, you will not be able to get your heels together.

Both flat feet and pigeon toes are usually due to the angulation at the knees which is very difficult to realign. Nevertheless it is wiser to attempt knee straightening by exercise than to modify the shoes, or add inserts, which only serve to disguise the true problem.

The importance of postural appearance has increased markedly during the past thirty years. Divorce and separation have become commonplace during this period, and now, with defacto relationships increasing, remarriage or a change of partner in later life is frequent. There is now a much greater need to retain physical attraction into middle and older life and with the earlier demise of the male, the pressure for this need is greater on the female. The last fifty years have seen the disappearance of two activities, which did so much to maintain good posture in women, namely praying

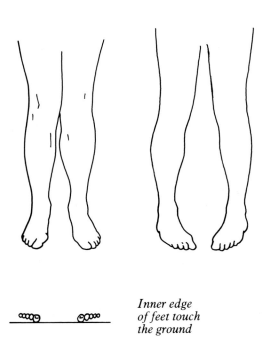

*Inner edge
of feet touch
the ground*

Figure 11-10. The knock-kneed person (a) before making any adjustments of the foot, ankle and knees for being knock-kneed and (b) after arching the foot, turning them in, locking the knees back.

and scrubbing. Praying was an excellent exercise for the lower back and the buttocks, particularly if the arms were unsupported. Scrubbing the floors exercised the whole of the back, but particularly the upper back and was an excellent exercise for breast development and firmness.

It is generally accepted that most of the problems referred to in this chapter are part of the ageing process and must, therefore, either be accepted or treated with cosmetic surgery. Their basic cause is, however, poor posture accentuated by gravity acting over a lengthy time span.

Early postural therapy will prevent most of these problems and, if they are already established, appropriate therapy will do much to correct them. When cosmetic surgery is necessary subsequent postural therapy is vital to prevent the recurrence of any disfigurement.

MENTAL

The relationship of the mind and posture is a two way process. The mind can affect the posture, and posture can affect the mind.

For centuries it has been recognised that the mind affects the body. Nearly two thousand years ago, Claudius Galin, the Greek, talked about different types of personality and their susceptibility to illness. The way the mind, particularly under stress and tension, affects the body has acquired a title — psychosomatic medicine.

The trouble with a psychosomatic diagnosis is that sometimes it is just too glib and simple to make. In fact, there is a constant interplay between mind and body, which cannot be separated, and which often makes diagnosis difficult. Has depression produced round shoulders, causing pain in the back? Or, have round shoulders caused a constant pain in the back, inducing depression?

It is accepted that the commonest cause of convulsions in children is infection, i.e. the body affecting the mind. Yet, when less severe mental disturbances such as sleep-walking or nightmares occur, a physical cause is not even considered. The effects of the body on the mind have taken second place to the effects of the mind on the body. This is a pity because it is frequently much easier to improve the body than to improve the mental attitude.

It would seem wise, therefore, to dignify the reverse of psychomatic illness, i.e. the body on the mind effect, with a name, to redress this imbalance.

> **I suggest SOMAPSYCHOTIC ILLNESS**

Some illnesses have a direct effect on the mind. Hepatitis (yellow jaundice) is so constant in causing depression that we speak of a jaundiced outlook, or being liverish. The constant pain of osteoarthritis can make a person bad tempered and morose.

However, the effects of the body on the mind may also be positive. Regular joggers, almost without exception, claim a mood elevation. In fact, any hard physical exertion is likely to be followed by a feeling of elation and of reduced mental tension. Opium like substances produced by the brain dur-

ing physical exertion have been discovered, and it is thought that these produce the mood elevation. However, there may be other factors working. Exercise may use up the circulating adrenalin which has been produced by the stresses of the day and, by increasing the blood flow to the brain, may increase the rate of recharging of the brain, so that it reaches a new threshold level of charge (Appendix).

Most cosmetic surgery is related to the effect of the body on the mind. A big, unsightly nose, a hare lip, a large birth mark on the face, overlarge or under-developed breasts, are all examples of how physical shape and appearance react on the mind producing shyness, lack of confidence or perhaps aggression. Unfortunately, by the time surgery is performed the mental effect has often become permanent. The psychosomatic effect of mood is often reflected in posture. A worried, sad, or depressed person may be literally "weighed down with problems" and stoop. When the sun shines and the world seems good, people "face up to life" and stand taller.

Unfortunately, as cosmetic surgeons have found, the appearance may be corrected by postural therapy but the psychological change is often deeply entrenched and, although the shoulders may be straightened, the personality may remain stooped. Yet, there is an advantage in correcting such posture. People react more pleasantly and with greater respect to a person who stands and sits with balance and composure and who moves with grace.

In girls, especially the body on mind and the mind on body effects can be seen acting in the one person. If the breasts are either much too large or too small, they produce self-consciousness (somapsychotic) and the self-consciousness makes them round the shoulders and stoop to try to hide the breasts (psychosomatic).

Where there is any obvious or imagined disfigurement, the owner will try to hide it, turning the good side of the face towards the world. This will distort the whole posture. A teenager with round shoulders and "winged scapula" (protruding shoulder blades) explained that she did not like her wide shoulders and tried to make them smaller. Very tall girls frequently ruin their posture by slouching, in an attempt to get down to normal height.

Short males, as well as females, try to gain

height with high heels. These, of course, make their poor posture even worse, and so reduce their true height. Among the short, stocky, sway-backed type of person, we often find an aggressive personality with a tendency to dominate. These people often become leaders in politics or business. Many of the dictators in history were of this physical type.

It is usually suggested that this tendency to dominate is a compensation for the inferior feelings they have acquired because of their short stature, but the clue to their behaviour is in their childhood and their early relationships with other children.

The group behaviour of children in the five to ten age group gives some clues. The leader of the small playground group is often short and stocky. This type of child is comparatively strong, agile and well co-ordinated, with movements that are balanced and controlled, and is capable of dominating the age group by physical superiority. He may be short, but he is likely to have a wide angle of obesity, denoting power and physical strength — and he matures physically early in life.

By contrast, the taller members of the group are often ungainly, clumsy, and have poor physical co-ordination and agility for their age. Their muscular strength is not, at this age, capable of controlling their limbs adequately. So they are easily dominated by their shorter classmates. It would appear that superior strength at an early age rather than inferior feelings produces the aggressive leader.

> **No matter what abnormality of posture exists, whether psychosomatic or somapsychotic, correcting will improve personality with more certainty and more quickly than would the exclusively psychological approach.**

TREATMENT: See Chapter 21 For

(A)   Essential Exercises 1, 2, 3.
(B)   Recommended Exercises 5-10.
IN ADDITION:
(C)   For round shoulders, exercise 4.
(D)   For sway backed, Exercises 22, 23.

CHAPTER TWELVE
POSTURE IN OSTEOARTHRITIS

*"He preacheth patience
that never knew pain."*

Proverb.

There is a quite simple explanation of the cause of osteoarthritis — and a simple remedy that works.

The U.S. Arthritis Foundation has estimated that 360 million people suffer from some form of arthritis — ten per cent of the world's population. It is probable that the current figure, taking into account developing arthritis, those who do not seek treatment, and those who have been misdiagnosed, is nearer 1,000 million.

Arthritis simply means inflammation of a joint, and while there are many different forms of the disease, osteoarthritis accounts for a high percentage of all cases. Osteoarthritis is one of the greatest scourges afflicting mankind, causing untold misery, and probably more pain over a longer period, than any other disease, including cancer.

Time and again I have listened to stories of pain, of restricted lives, of fear of sex, all resulting from osteoarthritis or back pain. And I have seen the joy when these lives have been returned to normal.

Why, if the answer is so simple, has it eluded the researchers for so long? Unfortunately thinking has generally been in terms of the man-made machine instead of the living machine. (Chapter 2)

In this age of specialisation, scientists have been looking at the trees instead of the woods. They have been looking, with their electron microscopes at the cell in the cartilage in the joint, or the chemistry within that cell, instead of the whole human being.

Diabetes may cause blindness, but the answer is to be found in the gland which produces insulin, not in the eye. The skin may turn yellow in jaundice, but the answer lies in the damaged liver or the bile duct blocked by stones, not in detailed examination of the skin. And the answer to osteorthritis lies not in the joint, but in muscular imbalance, and the multitude of things which can produce it — from deafness to ingrowing toe-nails.

In osteoarthritis, none of the changes in the blood occurs which one would expect with an inflammatory type of illness. In fact, the view that the condition is not caused by inflammation is being accepted and the condition is now often called osteoarthrosis (a changed condition of the joint) rather than osteoarthritis, (inflammation of the joint).

One must look elsewhere for the cause. The correct direction lies in our way of life. But the "experts", stating that the cause is unknown, and that there is no cure, persist in describing the disease in "man-made machinery" language. The text books and the journals state that osteoarthritis is a degenerative, wear-and-tear type of arthritis that progresses with the use of the joint and with age.

> **There is NO PROOF that osteoarthritis is caused by wear-and-tear, NO PROOF that it progresses with the "correct" use of the joint (only the "incorrect" use of the joint), and NO PROOF that it progresses with age (unless the "incorrect" use is allowed to continue).**

The terms degenerative disease and wear-and-tear are quite incompatible because degenerative is a term which implies disuse (wasting), whilst wear-and-tear means over-use. It would be hard to imagine a disease caused by both disuse and over-use.

The living machine improves in strength, efficiency and performance in response to correct stimuli. Joints and bones are no exception to this response. In fact, Wolff's Law stated in 1868 "that every change in function of a bone is followed by certain definite changes in its internal architecture and its external conformation". Civilisation has produced changes in the functions we give our bones and these changes in turn produce changes in both "the internal architecture and external conformation" of the bones making up the joint — i.e. OSTEOARTHRITIS.

If we return to the normal function of the bones and joints as designed by evolution the joints will gradually return to a normal, healthy, pain free condition — AND AT ANY AGE.

Following these principles for the past ten years, I have developed a programme specially designed to realign joints to the balanced position that evolution designed for them over millions of years. This realignment produces the correct pressure stimulus that is necessary to restore the osteoarthritic joint to a normal condition. The success of the programme has been repeatedly confirmed by patients' freedom from pain, return of joint mobility, and X-ray evidence.

The clue to osteoarthritis has been staring out of the pages of the textbooks on orthopaedics for years. There is a condition, (slipped upper femoral epiphysis) which occurs in teenage, in which the head of the femur sheers away from the neck of the femur, in the hip joint. Because of the pressure of the weight downwards and the pull of the muscles upwards, the head is always displaced backwards and downwards. The textbooks tell us that "if severe displacement is allowed to remain uncorrected, osteoarthritis ALWAYS develops in later life" or "osteoarthritis is an INEVITABLE sequel to cases in which marked displacement persists" or "when displacement is severe, the position cannot be accepted because of the CERTAINTY that painful osteoarthritis WILL develop in adult life".

The usual treatment is to remove a wedge of bone from the shaft of the femur (thighbone) to bring the head of the femur "once more into the weight-transmitting segment of the hip joint". The statement is made that "It often helps even when osteoarthritic changes are beginning to appear."

Although the development of good posture, particularly in heavy children with a large angle of obesity, might help to prevent this condition, once the condition has occurred exercise is contra-indicated and operation is invariably essential.

The condition demonstrates that if the head of the femur is not in its correct weight bearing place in the joint osteoarthritis WILL develop, and if the head is correctly realigned it WILL be prevented.

Except in the case of slipped upper femoral epiphysis, I prefer to prevent and correct osteo-arthritis by shoe modification and exercise, rather than by surgery.

The two bones comprising a joint must be compressed together adequately in order to stimulate growth of the cartilage covering the bone, and growth of the ends of the bone itself. As well, the bones must be properly aligned and able to go through their full range of movement.

At every joint there are muscles which pull the joint in one direction, and muscles which pull it in the opposite direction. The combined pull of the tension in these muscles compresses the bones against each other in the joint, stimulating the normal structure and health of the joint.

In the joints that bear weight, there is the additional compression that comes from the weight of the body.

> **The alignment of the bones in all joints comes from the relative tension of the muscles acting on the joints.**

The ability to move a joint through its full range depends upon the capsule of the joint (which is like a glove fitting around the joint) and the ligaments (strong bands of tissue which support and protect the joint), as well as the relative tension of the muscles acting on the joint. If the joint is not moved through its full range for long periods, sometimes for many years, the capsule and the ligaments shorten, preventing full movement. This shortening can be corrected at any age by exercising in the opposite direction to the shortening.

The riddle of osteoarthritis lies not in the joints, or even in the bones themselves. It lies in the muscles. And the treatment lies not in surgery or drugs, but in correct exercise therapy and the removal of heels from shoes, in order to align the joint as nature intended. OSTEOARTHRITIS COULD BE CORRECTLY CLASSIFIED AS A "MUSCLE IMBALANCE DISEASE".

If the wear-and-tear theory were true we would expect to find weightlifters, long distance runners, and ballet dancers crippled with osteo-arthritis at a very early age, possibly even before they left school. Yet these people maintain healthy and supple joints for much longer than the general public. Competitive weightlifters in full training handle in excess of 15 tons of weight in a variety of lifts in a training session. Even after ten years of such training six days a week, weightlifters rarely suffer from osteoarthritis and are in fact just reaching a peak of performance. Although the movements involved are often jarring and stressful, it is common to see veteran weightlifters with excellent muscle tone and posture still training with great flexibility and freedom in their joints while many of their non exercising counterparts are distorted with crippling osteoarthritis.

The most dramatic and demanding race in all sport is the marathon. It has so captured the imagination of the public that it is common to see thousands of competitors in major races. These competitors along with many millions of joggers world wide train over distances that would have alarmed doctors 20 years ago. In spite of the

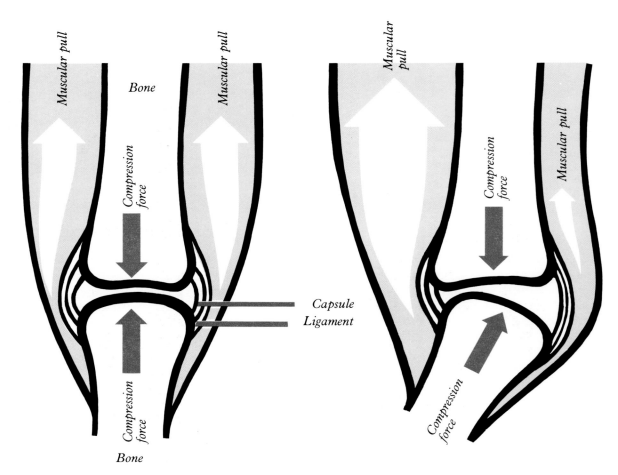

Figure 12-1. The bones of a joint need to be pressed together to stimulate the maintenance of its health. This compression comes from body weight and/or the pull of the muscles acting on a joint.

Figure 12-2. The alignment of the joint comes from the relative tension of the muscles acting on the joint, and should be balanced to achieve correct alignment and compression of the joint.

pounding their weight-bearing joints are supposed to receive, osteoarthritis is not a significant problem to them. It is probable that what little joint damage they suffer is due to the recent general introduction of heels to road shoes.

But of all sportswomen and men, ballet dancers subject their joints to the greatest volume of stress. As professionals, their activity lasts five or six hours a day, six days a week. The stress on a ballet dancer's joints is greater than that of any other activity. As Margot Fonteyn produces a very crescendo of emotion inside us, moving with evolution's perfection at the age of sixty, who can imagine arthritis beneath those youthful limbs?

The world famous choreographer Ninette de Valois, goes on and on, flexing and bending in her eighties. Postural excellence and continuous usage!

The concept of tear (i.e. injury) to a human body is indisputable, but the idea of wear as connected to living material must be discarded. The Persian carpet may wear, like the car's piston, but the wool on the sheep's back does not wear out — it regenerates continuously. The rubber, steel or plastic tyre may wear out, but the skin on our feet does not. In fact, it gets thinner if it is NOT used.

Wear (pressure stimulus) does not produce degeneration in the joint. However, where this pressure is exerted on a part of the joint not de-

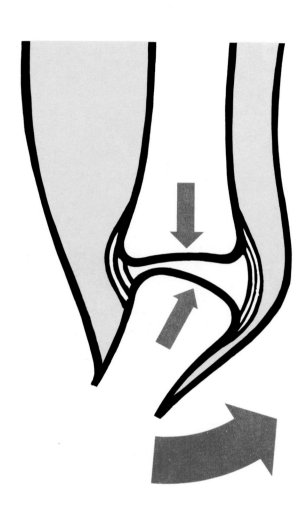

Figure 12-3. *Capsules and ligaments shorten on the side of the stronger muscles. Exercise is necessary in the opposite direction (i.e. of arrow).*

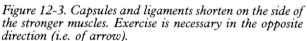

Figure 12-4. *Osteophytes (red) appearing at the knee, when its alignment is distorted by being knock-kneed.*

signed to take it, as happens when the joint is not correctly aligned, the body thickens the bone and even lays down bone where none previously existed. These bony bumps known as 'osteophytes' are a sign of osteoarthritis. Their obvious job of bearing weight or pressure, caused by the adaption of bone structure to distorted function, seems self explanatory.

X-rays confirm that osteophytes occur in that part of the joint bearing weight, when the joint is out of its correct alignment. The build-up of bone happens in the same way as the thickening of skin on the soles of bare feet or the skin on the tips of the fingers of a violinist.

On the other hand degeneration occurs in the area of the joint where normal pressure stimulus is reduced, and here, the cartilage becomes thin or absent, and the underlying bone weakens as its structure is altered.

In the osteoarthritic joint, we have weakening of the bone structure, where the pressure has been reduced from that which nature has designed, and strengthened areas of bone or even the laying down of new bone, where pressure has been increased above nature's design. The consequence of this is

PAIN AND DISTORTED JOINTS.

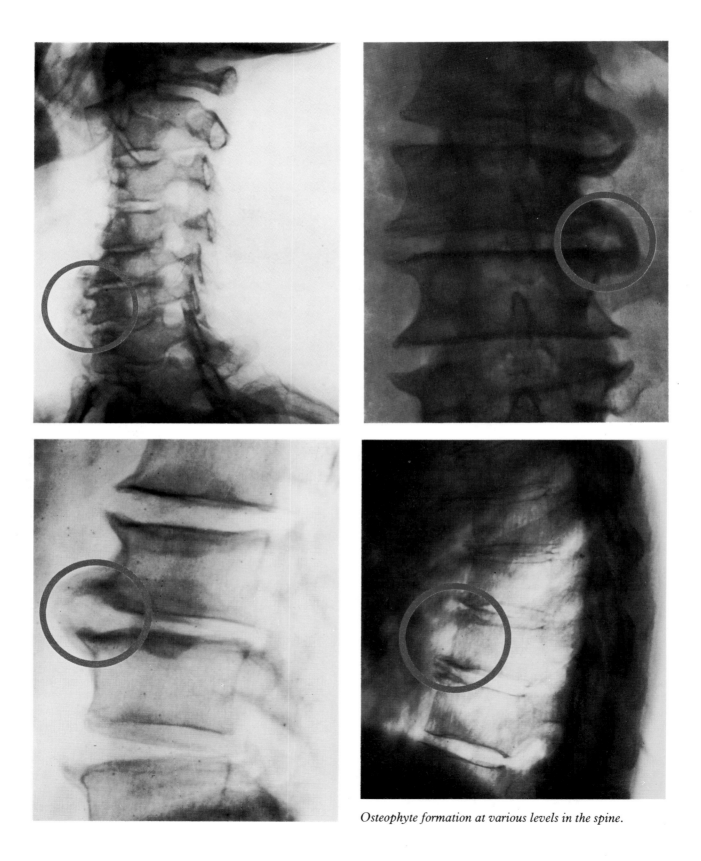

Osteophyte formation at various levels in the spine.

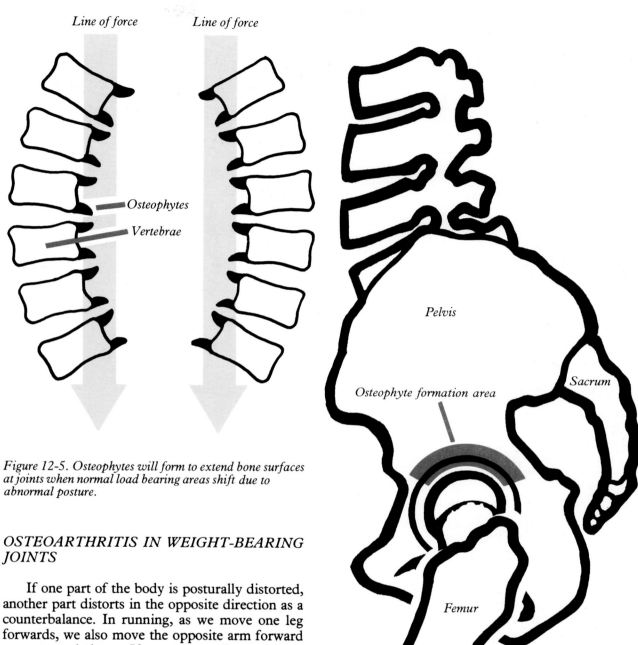

Line of force   *Line of force*

Osteophytes
Vertebrae

Figure 12-5. Osteophytes will form to extend bone surfaces at joints when normal load bearing areas shift due to abnormal posture.

Pelvis

Sacrum

Osteophyte formation area

Femur

Figure 12-6. Osteophytes at hip joint.

OSTEOARTHRITIS IN WEIGHT-BEARING JOINTS

If one part of the body is posturally distorted, another part distorts in the opposite direction as a counterbalance. In running, as we move one leg forwards, we also move the opposite arm forward as a counterbalance. If we start to slip, we throw ourselves in the opposite direction to try and keep our balance.

If one weight bearing joint (ankles, knees, hips and joints of the spine) is thrown out of alignment, at least one other joint must be misaligned, and in practice usually all the weight bearing joints need correction. Those who develop severe osteo-arthritis in one weight-bearing joint, sooner or later develop it in the others.

OSTEOARTHRITIS OF THE HIP

We have seen that people who have an excessive curve in the small of the back stand with their hips bent. If these people wear heels, the hip flexion is even greater. With 8cm heels, it can be as much as twenty degrees.

The hip is a simple ball and socket joint. When it flexes forward the socket rotates forwards and the ball rotates backwards. The back of the socket and the front of the ball are now taking most of the weight, and are, therefore, receiving an increased amount of stimulus.

And because of the increased curves of the body, the centre of gravity has moved further forward in front of the hip joint. Thus the weight (W) is borne on the front of the rotated joint, and has become a shearing force (S) as well as a compressing force. If we resolve this force (W) it can be seen that the force that compresses the joint is reduced to C. The tangential force (S) which tends to force the head of the femur out of the back of the joint, also

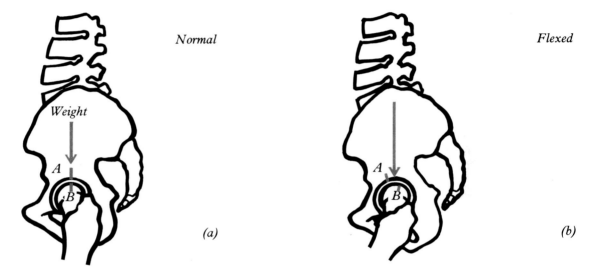

Figure 12-7. When the hip is in a flexed position (b), the back of the socket and the front of the ball take on increased weight (i.e. stimulus) and the areas designed for weight bearing (i.e. at A and B) bear a reduced weight.

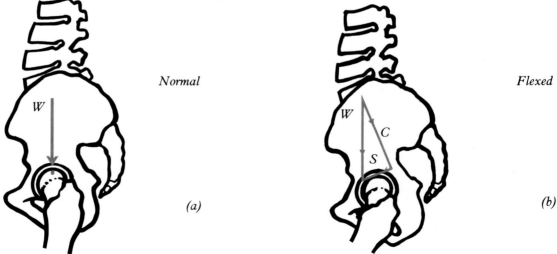

Figure 12-8. When the hip is in a flexed position (b), the compressing force W is reduced to C, and a shearing force S is produced.

tends to collapse the joint even further, and has to be resisted by the action of the buttock and hamstring muscles in order to prevent this collapse.

The sexually determined difference in the shape of the bones of the hip plays an important part in the production of osteoarthritis. The thigh bone has a short almost horizontal piece which fits into the hip joint and a long vertical shaft which ends at the knee. In women the short piece (the neck of the femur) is horizontal and comes out sideways from the hip. In men it is angled some-

what backwards from the joint as well as sideways, tending to also angle downwards. Whether this is due to the actual shape of the male femur or the slope of the male pelvis, the result is the same. When there is bending at the hip the thigh bone of the male moves out sideways, whilst the female's bends straight forwards.

It is anatomy, and not modesty, which keeps the female's knees together when she bends.

When a male is bent at the hip (due to heels on his shoes and a lordotic, sway-backed posture) he

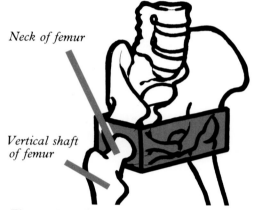

Neck of femur

Vertical shaft of femur

Figure 12-9. A schematic diagram of the pelvis, to show the region represented in figure 12-10 (a) and (b).

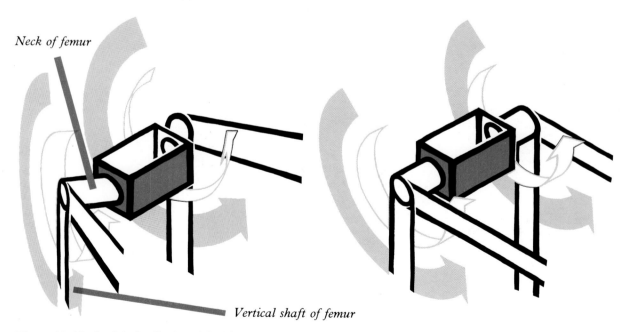

Neck of femur

Vertical shaft of femur

Figure 12-10a (male). On flexion of the hip, the thigh bone goes out sideways as well as forwards.

Figure 12-10b (female). On flexion of the hip, the thigh bone bends straight forwards.

pulls his thighs inwards towards the midline by the contraction of the muscles on the inner side of his thighs (the adductors), in order to keep his weight over his knees. When this happens the head of the femur (the ball) tends to move backwards and sideways creating further displacement of the head of the femur from its intended position in the joint. For this reason men have always preferred lower heels to women, and have a greater tendency to osteoarthritis of the hip. However, when women wear high heels, they increase the likelihood of osteoarthritis developing.

The net effect of a sway backed posture with its resultant bent hips and knees is that:
(1) The compression force stimulating correct structure has been reduced.
(2) The compression force is acting in front of its correct line.
(3) A shearing force has been introduced, tending to force the head of the femur sideways and backwards out of the joint.
(4) The wrong parts of the ball and the socket are load-bearing.

The consequences are:
(1) The reduction in the compression force on the correct areas produces degeneration through disuse.
(2) The new areas subjected to additional load-bearing and shearing forces produce abnormal strengthening of bone and the formation of new bone where none existed (osteophytes).

My original concern in treatment (Chapter 21) was whether I could persuade patients to do the necessary exercise. This, in fact, proved to be no problem. Relief of pain was a wonderful motive for action and relief normally came within fourteen days.

The pain of osteoarthritis is considered to be due to stagnation of blood in the bone, and the correct specific exercise therapy quickly reduces this stagnation.

For treatment to be effective, it is absolutely essential that a person should wear shoes without heels, and if one leg is shorter than the other, the shoe of the shorter leg must be built up along its full length. If the heel only is raised the knee becomes bent forward, increasing the possibility of osteoarthritis of the knee and doing nothing to compress and stimulate the hip joint.

With severe osteoarthritis of the hip, the joint starts to collapse, so that the affected leg becomes shorter than the other.

If the shoe of the shorter leg is not built up to the correct height, then stimulus of the bone at the joint will not take place. This leg length must be checked regularly, for as the hip joint regenerates, the shortening will disappear, and the build-up of the shoe will have to be removed.

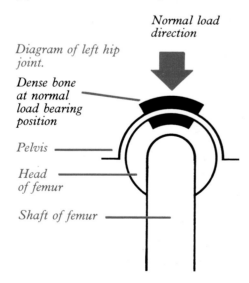

Figure 12-11a. Hip joint. Good posture.

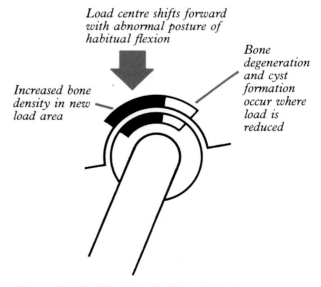

Figure 12-11b. Hip joint, held in flexion by poor posture. Degeneration occurs in some areas, strengthening and osteophyte formation occurs in other parts.

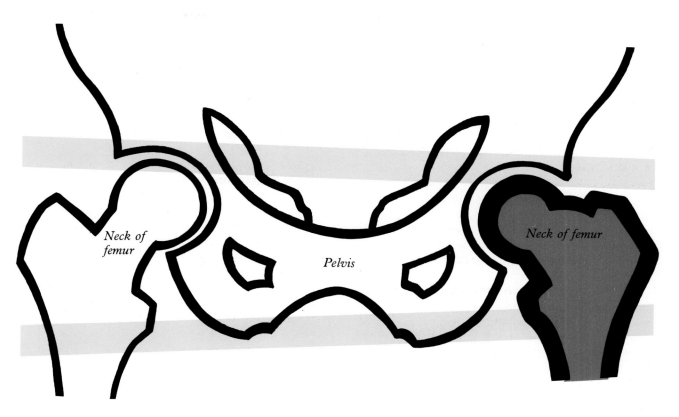

Figure 12-12. In severe osteoarthritis of the hip, the joint collapses, mainly by flattening of the neck, making the affected leg shorter.

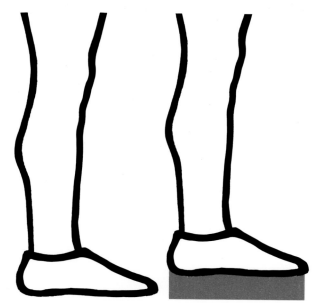

Figure 12-13. The shoe of a shorter leg must be built up along its whole length, in order that the hip joint is properly stimulated by pressure.

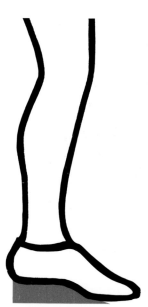

Figure 12-14. When the heel only is raised, the knee becomes bent. This will tend to produce osteoarthritis of the knee. The leg becomes fore-shortened, so that the hip joint is still not stimulated.

> **A seventy year old patient regained just under two centimetres in four months, by building up the whole shoe and doing appropriate exercises. The measurements were taken accurately by X-ray (scanogram) both before and afterwards.**

There are many difficulties in trying to measure the length of the legs, and this should be left to a doctor fully conversant with all the traps. For instance, a knee which will not straighten fully, will give a shortened reading for the length of the leg (if it is bent forward four inches the leg will seem one inch shorter, and this should be treated rather than providing a built-up shoe). In the very obese, measurement becomes impossible, because the bony points cannot be felt properly and the only accurate measurement is by X-ray (scanogram). If you check the feet with the patient lying flat, one leg may appear longer than the other, but this can be because the pelvis is tilted to one side or the other. However, if the patient sits forward, with the knees equally straight, much of any pelvic tilt may disappear, and the position of the feet gives a more accurate indication of leg shortening, even in the obese.

Pelvic tilt is very important, for it produces apparent (false) shortening of the leg and requires exercise to correct. Whereas true shortening requires a build-up of the whole length of the shoe, pelvic tilt is usually the result of muscular imbalance in one leg (adductors stronger than abductors). The logical correction of this deformity is exercise to strengthen the muscles which pull the leg out sideways (abductors) not the cutting of the muscles which pull in the opposite direction (adductors) or which flex the hip (iliopsoas) as is sometimes the practice.

> **Osteoarthritis is the result of imbalance of muscle tone, and not an inherent disease of the joint.**

The problems of back trouble and osteoarthritis in weight-bearing joints would seem to appear more frequently and earlier in life, when one leg is shorter than the other, whether this is a true shortening or a tilt of the pelvis. This is to be expected. If a building or a crane were tilted in this way, it would be disastrous. Yet, rarely are leg lengths measured — and, unless the difference is at least half an inch, it is most unusual for anything to be done. However, when there is a discrepancy, the weight, and therefore the stimulus, is borne largely by the shorter leg, for most of the day. The shorter leg appears to make the major effort in walking, with the longer leg swinging out sideways. It is usual to find that the muscles of the shorter leg are thicker and more developed.

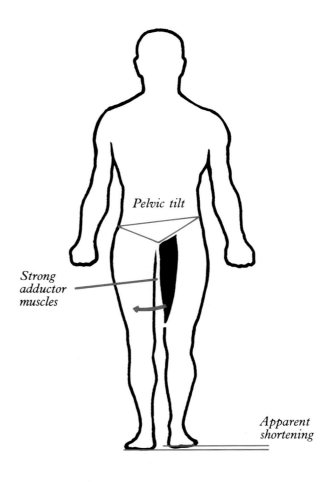

Figure 12-15. Tight adductor muscles in one leg produce pelvic tilt, and apparent shortening in this leg, and apparent lengthening of the opposite leg.

The muscles which pull the leg across the midline develop excessively when heels are worn (particularly by males), and although this produces no tilting of the pelvis it does produce muscular imbalance with the adductors stronger than the abductors and the flexors of the hip stronger than the extensors. The resulting misalignment of the bones of the hip joint is likely to produce osteoarthritis.

Many other factors besides heels produce persistent flexion at the hip joint. Car, motorcycle, bicycle or horse back riding are all examples.

Professional squash players seem particularly

Figure 12-16. Persistent flexion of hip produced by car, motor-cycle, bicycle and horseback riding.

prone to developing osteoarthritis of the hips, knees and back at a very early age. Jarring and wear and tear are the explanations given, but the constant semisquat position the game demands is almost certainly the culprit. It is a pity that these people cut short their careers because of a failure to practice correct preventive exercise therapy (Chapter 21).

Modern civilised workers start their day sitting at the breakfast table. They transfer to a car seat, then to an office, factory or tractor seat, back to a car seat and end their day sitting watching television and/or drinking.

Then if the knee is bent for most of the day the muscles, tendons, ligaments and joint capsule at the back of the knee thicken and at the front thin and stretch, preventing the normal straightening of the knee when standing.

> **The chairborne life is a major factor responsible for ostearthritis in all weight-bearing joints.**

The male is again worst affected, because of the anatomical differences between the sexes at the hip joint. Heels on shoes also cause the knees to bend. The compression force on the joint is reduced and acts on the wrong part of the joint, and body weight tends to collapse the knees even further. As in hips, there is weakening of the bone and cartilage where the normal force has been reduced, and strengthening, including the production of lipping (osteophytes), in the parts of the joint where increased force has occurred.

In diseases such as ankylosing spondylitis and rheumatoid arthritis, both the hips and knees become bent, In both these conditions osteoarthritis commonly occurs in these joints.

Prolonged weightlessness as in space flight could also produce problems. Exercises have been designed to maintain muscle tone and circulation, but bones become reduced in structure and strength because they are receiving no stimulus from gravity. Although there is nothing to disturb the alignment of the joints, the reduced stimulus to joints is likely to produce osteoarthritis if astronauts are long in space.

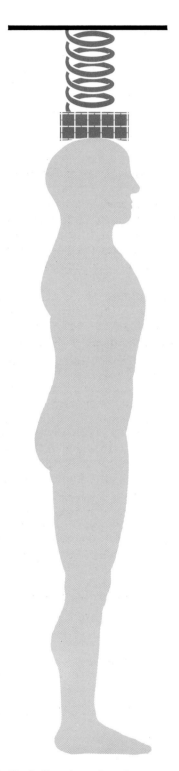

The other weight bearing joints are those of the spine and ankle and the same principles apply to them. The low incidence of osteoarthritis in ankle joints is probably due to the normal alignment of the ankle joint while we are seated. The joints of the spine are dealt with in depth in Chapter 13.

In joints that do not bear weight, it is the pull of the muscle tone (tension) alone, which stimulates the health of the joint.

If you have a job that requires you to bend your wrist forward continuously, the muscles responsible will develop and their tone will increase. The joint will be pulled to the side of the stronger muscles, even when they are relaxed.

This side of the joint will have greater compression than the other establishing conditions conducive to the development of osteoarthritis.

A simple example is the end joint of the fingers of the typist. In time they may become almost fixed in a bent position, so that very little or no movement can take place at the joint. Painful arthritic lumps then appear on the backs of the joint.

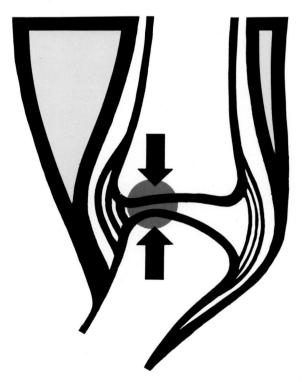

Figure 12-17, Coil spring of 100kg pressure, for astronaut of 90 kg weight, to maintain bone and joint strength.

Figure 12-18. Greater compression on flexed side of joints.

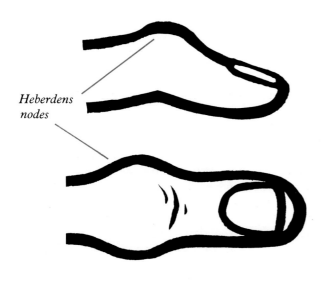

Heberdens
nodes

Figure 12-19. Painful arthritic lumps (Heberdens Nodes) over the end joint of the fingers.

Occupation is normally the vital factor governing muscle tone. A life-time spent bending forward and looking down, as a carpenter may do in sawing, measuring, and nailing, will cause the muscles in the front of the body to become developed and the person to become round shouldered and stooped. The house painter, who will at times be reaching up and at other times be bending down and forward, is likely to maintain good posture. The gnarled, knobbly painful arthritic fingers of the elderly woman are the result of activities that have changed the balance of her muscle tone. During her life, the housewife has peeled potatoes, wrung out washing, polished, scrubbed, weeded the garden, knitted, and sewn — every one of which required bending of the fingers, and none of which required that the fingers be straightened. It is not surprising that osteoarthritis develops in her fingers and wrists.

The perfect example of correct and specific exercise therapy occurred when I treated a man who, for thirty years, had worked in a shoe factory, pushing a sheet of leather into a cutting machine with his left hand (straightening the fingers), and pulling a lever with his right hand (bending the fingers) to cut out the pattern. He suffered osteoarthritis in both hands.

> **By exercising the muscles that bent the left hand and the muscles that straightened the right hand (the opposite to his normal use) the years were rolled back and his symptoms disappeared.**

TREATMENT: (for details see Chapter 21)

The treatment of *weight bearing joints* is fairly straightforward. It is the correction of posture, and usually requires exercises to strengthen the muscles that straighten the knee and hip joints, and the muscles which restore the normal curves to the spine, particularly the neck.

With *non-weight bearing joints* one can decide on the muscles which require strengthening, by the history of previous use of the joint, and by the restriction in the movement of the joint. Where a joint will not straighten fully, the muscle that straightens the joint must be exercised. If the elbow will not straighten fully, the triceps muscle which straightens the elbow must be exercised.

An elderly man whom I treated had an arm which was permanently bent about 30 degrees at the elbow as a result of it having been broken three times. He had been told that there was nothing that could be done for it. Naturally it was a nuisance and it was also painful for osteoarthritis had developed. A few months of exercising the triceps muscle straightened his arm, and ended the pain. He was left with a web of loose skin hanging down from the elbow where it had been stretched for so many years.

> **It is essential to select the correct muscles to exercise. If you further strengthen the stronger muscles, the condition will become worse. It is also vital that each exercise is performed correctly.**

The problem is not only one of making the weaker muscle pull against a stronger one, (this is not difficult since the brain not only sends messages to the weak muscle to contract, but at the same time sends messages to the stronger muscle to relax), but also of pulling against shortened and strengthened ligaments and joint capsules. Ligaments are supporting bands around the joints and capsules are glove like coverings of the joint. Over the years, these have become shortened and strengthened on the side of the joint with the strong muscles, and stretched and weakened on the opposite side. The shortened thickened tissues must be stretched by the increased tension developed as weaker muscles are gradually strengthened. This takes time.

Improvement is usually rapid in the weak, for the correction is of minor muscle imbalance and the stretching of comparatively weak capsules and ligaments. If the pull in one direction is seven kilos and five kilos in the opposite direction — the weak muscle has only to improve its strength by two kilos. However, if the pull of one is two hundred kilos, and the other is one hundred kilos, there is one hundred kilos of muscle strength to make up and this will take time.

The elderly and the weak can take heart for their problem is not so difficult.

It is understandable that the first attempts at exercise may be difficult, though rarely painful. The difficulty is especially marked in the finger joints, where any movement at all in the joint nearest the finger-nail may seem impossible. However, with perserverance it starts to move, and then progress is rapid.

Patients rarely need much encouragement to exercise. A frequent comment is "Thank God you are not giving me tablets, Doctor". The relief of pain is a major incentive. If exercising stops during the early part of treatment, the pain soon returns and patients get back to exercising smartly. Many decide to continue the exercises indefinitely, even though this is not necessary in many cases. (See Chapter 21.)

CASE HISTORIES

A 47 year old woman who had suffered arthritis of the knees for years, and who wept silently at night with the pain, was running pain free within a month and playing badminton, within two months.

A 42 year old man with severe osteoarthritis of the hip had been on the waiting list for an artificial hip joint for two years. To dress he had to step into his trousers on the floor and have his wife pull them up and put on his socks and shoes. Within two months he was not only dressing himself, but playing bowls and golf. Nine years later he is still pain free.

An ex-commonwealth weightlifting champion had considerable muscle imbalance, affecting several joints. He had competed in the days when the legs were split, one forwards one back, in lifting the weight, rather than the present practice of squatting. He had always planted his right leg forwards and so had a lopsided development. Twenty years previously x-rays showed such bad osteoarthritis of the knees that he was forbidden to lift again. He had subsequently developed arthritis in his hips and spine. He attended a gymnasium with first class equipment and the pain, limp and much of the wasting in the left leg and buttocks disappeared in under three months.

Osteoarthritis is a result of our way of life. Such things as drugs, acupuncture or manipulation do not alter our way of life and are, therefore, unlikely to offer more than temporary relief from pain.

The proponents of spare parts surgery might claim rapid relief from symptoms by putting in an artificial joint. The risks of such surgery are considerable and the surgery does nothing to improve the other weight bearing joints. It may in fact, increase the probability of osteoarthritis in these other joints. Artificial joints are now being used to replace almost every joint in the body, except the spine. The method of treating the spine is to immobilise the joint with a bone graft. In spite of its popularity, 'artificial joints' is 'artificial treatment'.

Structural distortions can be reversed by treatment that naturally returns the body to normal load bearing function and gives quick relief from pain. I have successfully treated more than 300 osteoarthritic people, and know that the multitudes who suffer from this terrible afflication of the western world can also be helped if my methods are universally accepted. The proper application of the principles outlined in this book will, in fact PREVENT osteoarthritis from occurring.

Treatment: See Chapter 21
(A) In any weight-bearing joint.
  (i) Essential Exercises 1, 2, 3.
  (ii) Recommended Exercises 5-10.
  IN ADDITION:
  (iii) If round shouldered Exercise 4.
  (iv) If sway-backed Exercises 22, 23.
  (v) If pelvic tilt is also present Exercises 11, 12.
(B) In the shoulder joint Exercises 1, 2, 3, 4, 15.
(C) In the elbow joint Exercise 14.
(D) In the wrist Exercise 16.
(E) In the finger Exercises 17 and/or 18.
(F) In the toe or foot Exercises 19 and/or 20.

A WARNING

Although the treatment of osteoarthritis is simple, its diagnosis is not. Osteoarthritis must be diagnosed by a qualified medical practitioner. There are many forms of arthritis, and this treatment applies only to osteoarthritis. The correct diagnosis can only be made by a qualified physician for the symptoms of other diseases may mimic those of arthritis. Mrs. Brown down the road may have cured the ache in her hip by correcting her posture, but pain in the hip caused by cancer of the stomach or tuberculosis of the spine will not respond to exercise.

An even worse trap occurs when two diseases are present at the same time. A man whose posture obviously suggested osteoarthritis of the bones of the spine, also had pain in the neck with pain and locking of the left knee, all of which could have been due to osteoarthritis. Furthur investigations showed a tumour of the spinal cord itself, and at the time of operation it was felt even a month's delay could have produced permanent paralysis of all four limbs. The osteoarthritis of the spine, which was indeed also present, was treated a year later.

Rheumatoid arthritis is not osteoarthritis and does not respond to similar treatment. In the acute phases of the illness, exercise will almost certainly make the condition worse. Where the disease has appeared inactive over a period of years, some of the deformity of the joint may be improved by appropriate exercises. Under pressure, I have treated three cases of active rheumatoid arthritis with posture exercises and exercises to try to correct the deformities of the hands. One became much worse, two improved considerably. Much research would be required to evaluate any possible benefit from such exercises. But, certainly, many of the exercises sometimes prescribed are likely to make the condition worse.

CHAPTER THIRTEEN

POSTURE AND BACK PAIN

*"And when your back stops aching
and your hands begin to harden"*

— Rudyard Kipling (1865-1936)

YOUR BACK IS A CRANE!

A beautifully structured crane! A crane that would win any design award. It works in any direction, through a range of more than 180 degrees; it can lift anything from a feather to hundreds of kilograms, and it can be used in whole or part as required.

Yet, stuck on the walls of factories all around the world are posters warning that your back is not a crane. If you tell a lie or put forward a misconception often enough, it will be believed and the task of changing attitudes to backs is enormous.

Because this basic concept is misunderstood at least six million Americans suffer from back pain, and back problems cause approximately eight per cent of the total absence from work because of sickness in the United Kingdom, costing the country over 100 million pounds a year.

If trade unions and employers around the world were to live in perfect harmony and strikes became extinct the savings may not even equal the losses from back complaints.

No significant inroads into the immense problems of back complaints will be made until the correct basic principles of back function are accepted and understood. Difficult, indeed, when Health Departments insist on their posters that the back is not a crane!

> **The back is an excellent crane, with millions of years of the best research — the trial and error of evolution — behind it.**

It has a wonderful back-up service, for, like all goods manufactured by the Evolution Group of Companies, it has an inbuilt and automatic repair mechanism. Furthermore, it will modify its structure acccording to the work given to it, a refinement which no other manufacturer is offering. The makers, however, do insist upon two things:
(1) that the base is not put on a slope by raising the heels;

(2) that it is not left idle for long periods.

Provided these two conditions are honoured the manufacturer guarantees it indefinitely.

Your back uses a simple engineering principle, whereby a small force can be used to lift an enormous weight. It uses this principle at every joint along its length. It also uses the principle on all the joints of the spine together, or in combinations of just some of them, and in all directions — forwards, backwards, sideways, or in combination of these movements. What a crane! No man-made crane is capable of this.

In playing a snooker shot we may bend the spine forward in the lumbar region, twist it to the left in the upper lumbar region, rotate to the left in the thoracic region of the spine, and bend the spine back and rotate it to the right in the cervical region.

THE TWO POLE LIFTING MECHANISM USED BY THE SPINE

This lifting mechanism is illustrated in Fig. 13-1. Two poles, XY and YZ, are hinged at the point Y. The end Z is fixed to the ground and the end X bears the weight W. The lifting force is applied horizontally at the point Y. It requires no knowledge of mechanics to realise that when the poles are nearly vertical, then a small force F will lift a very heavy weight W. In fact, in theory, just as the vertical is being reached, an infinitely small force will lift an unlimited amount of weight. This is the mechanism of straightening any joint. And the greatest force is generated when the joint is almost straight.

On the other hand, it is very obvious that, when the poles are almost horizontal, an immense force would be needed to lift even a moderate weight.

So nature, in its magnificent back design, uses this mechanism only when the poles are nearly straight. Each joint in the spine is designed so that it can *only* move through a small range (3° to 8°); requiring only a small force to straighten the joint. Nature has designed a whole series of these joints (25) so that the total mobility is approximately 180°, and the combined lifting forces of each of these joints are co-ordinated along the spine. Comparatively small muscles straighten the spine, despite the fact that it can lift very heavy weights.

Nature's crane — our spine — is a series of

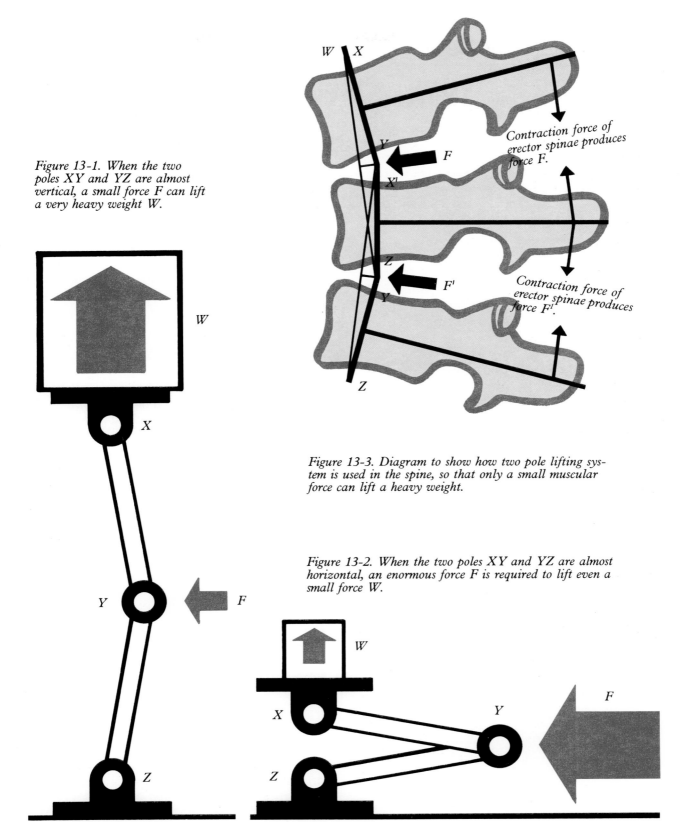

Figure 13-1. When the two poles XY and YZ are almost vertical, a small force F can lift a very heavy weight W.

Figure 13-3. Diagram to show how two pole lifting system is used in the spine, so that only a small muscular force can lift a heavy weight.

Figure 13-2. When the two poles XY and YZ are almost horizontal, an enormous force F is required to lift even a small force W.

slab-like bones placed on top of each other like a column, and separated by a disc, the infamous disc. There is a bony projection sticking out from the back of each bone to which is attached a muscle. When this muscle contracts it pulls adjacent bony projections towards each other, producing the force (F) to straighten adjacent bones of the spine.

Down the spine are a similar series of small muscles, which straighten the spine when it has been bent sideways using the same principle.

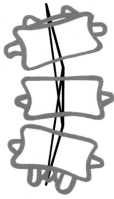

Figure 13-4. Similar use is made of the two pole system to straighten the spine, when it has been bent sideways.

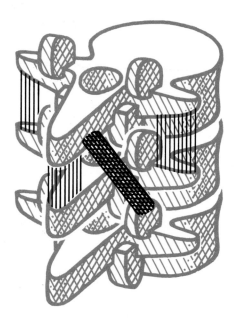

Figure 13-5. Muscles used to rotate the spine.

There are also sloping muscles attached to the spine, which rotate it around its length.

There are, also, long muscles of the back which control movements over many consecutive bones of the spine, with the smaller muscles helping to stabilise these major movements.

The result is a 'crane' which will bend in all directions over a considerable range, and capable of lifting very heavy weights. It is the same 'crane' that can do a double somersault with a twist and land safely in the upright position.

Mechanical cranes are clumsy and inefficient by comparison. If industry were to copy nature's crane and use nature's muscle design (see Appendix — Muscle Contraction), the resulting robot would efficiently lift both very heavy and very light weights, and its extreme delicacy, accuracy and flexibility of movement would make it invaluable for complex manipulations.

The British Back Pain Association and the New Zealand Accident Compensation Corporation both maintain that the bones of the spine and the discs are compressed when lifting with the back. These authorities claim that just holding a 50 kilogram weight, with the back horizontal, produces nearly enough compressive force to break down the bones and the discs, and that the compressive force is even greater when actual lifting takes place.

Yet weights of over 200 kilograms can be lifted in the stiff legged deadlift, in which only the muscles of the back and hip joint are used. The lifter bends down to grasp a barbell, keeping his legs straight, with knees locked. He then lifts the barbell up by straightening his back, until he is standing upright.

> **Do not believe the myth that your back is not a crane and should not be used for lifting.**

In actual fact, the two-pole lifting mechanism distends the spine (stretches it out), tending to seperate the bones and the disc. When the back is horizontal there is no compression of the spine whatsoever, and it is only as the spine approaches the vertical that compression occurs. In practice, when lifting heavy weights, it is only near the vertical that the lifter is aware of his spine being compressed.

There is another way in which your back can act as a crane, in which it closely resembles a man-made machine. When all the muscles of the spine — those that bend it forwards, backwards, and to either side — contract at the same time, the spine is held in a rigid position, known as isometric contraction, because the pull of one muscle is counter-acted by the pull of its opposite number, so no

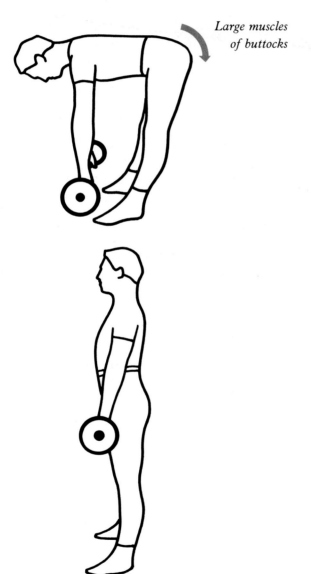

Large muscles of buttocks

Figure 13-6 and Figure 13-7. When the spine is held rigid, it can be raised and lowered by the muscles of the buttocks, which have to be large and powerful.

movement takes place. The spine, and the pelvis to which it is fixed, then becomes a straight, rigid level which can be lowered and raised by bending at the hip joint. This is not a very efficient system, so the muscles which straighten the hip joint — the muscles of the buttock — have to be big and powerful.

Whether in every day use to pick up a teaspoon, or in lifting very heavy weights, nature uses a combination of all available methods. The spine lifts by straightening the individual joints, and the hip joint is used to straighten the whole body. For good measure, the knee joint is also brought into play by bending, and then straightening and locking. Overseeing all, is the super computer (brain,) which brings all the necessary muscles into appropriate action, both in timing and strength, in sweet harmony. The simplicity and beautiful co-ordination of the human body, when acting as a crane, would bring ecstacy to a design engineer.

The designer of any man-made crane insists on the base of the crane being absolutely horizontal — even a fraction of a degree may upset the mechanics of lifting with the danger that the machine may topple over. We tilt the base of our crane by 15° to 50° when using raised heels and wonder why we have trouble.

The Health Department posters should read: "Your crane is not designed to have raised heels".

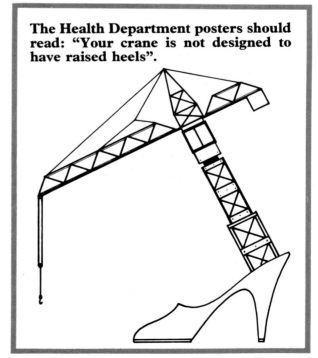

The "Experts" encourage further destabilisation of the base of our crane, by advising that one foot should be placed further back than the other, when lifting from the ground. As the knees bend, the heel of the rear foot comes off the ground, producing a week unstable base, particularly if the surface is slippery.

Figure 13-9. If one foot is placed behind the other, for lifting, the rear foot is unstable.

This is the traditional method of teaching nurses how to lift patients. No wonder back injury is the major disability of nurses, costing hospitals millions each year.

To maintain a healthy spine, we need muscles with good tone to hold the joints compressed against each other. The alignment of the joints must be held correctly by a balance of muscular tone between the bending muscles (flexors) and the straightening muscles (extensors). The joints must periodically move through their full range. In other words, we need *good posture*.

The other requirement of our crane is that it must not be left idle. Yet we are constantly told to do just that. We are advised to lift with our legs and not our backs.

If we do not use the muscles of our back, they will degenerate, and we will get back trouble.

We are told that we must lift like the weight lifters, who "keep their backs straight and lift with their legs". Here, I move into the field of the specialist, though not of the eminent, for I have spent over 40 years studying, researching, and occasionally practising the sport of weight lifting. My early love of mathematics attracted me to weight lifting and the mechanism of the body. This, in turn, led me to medicine and a joyful admiration of the workings of nature.

Nature has built the spine with 24 movable bones. It is both stupid and conceited to insist that these bones should not be moved and that the spine should act as a single rigid beam.

Professional ballet dancers rely for their livelihood on being free from back injury. Yet, everyday they practice an exercise at the bar, which would be condemned by any respectable authority on backs. And they do this hip bending exercise not only with straight, lock-backed knees, but also on the soles of their feet, or on points.

The truth is that all of us do and should use our backs in lifting, whether we bend our backs or endeavour to keep them straight with the isometric contraction of the spinal muscles.

> **You should not be afraid to use your back when lifting. Your muscles will thrive on the activity.**

Many back injuries are sustained by just bending down, without lifting anything. The explanation for this is simple. I have found when examining these people that usually, they either have feeble muscles throughout the length of the back, or else the muscles are extremely well developed in one region and almost non-existent in another. Their buttock muscles are often very poorly developed. The poor back and buttock muscle development is reminiscent of that of our ape ancestors, who are only partially and weakly designed for upright posture.

On the other hand, the rice planter in the paddyfields, after countless centuries of experiment and evolution, uses the back, not the knees, for bending as it is much more efficient to use the small muscles of the spine than the big muscles of the legs.

BACK PAIN AND THE CHAIRBORNE SOCIETY

> **The tone of all muscles must be good and the tone of opposing muscles must balance.**

The muscles that bend you forward must be as strong as those that bend you back, and those that bend you to one side must be as strong as those that bend you to the other.

The problem is that we spend most of our days in a chair or standing at a bench, and our back and buttock muscles waste away. Our chairs, workbenches, desks, cars and shoes all contribute to the ruin of the alignment of our posture, by their poor design. Nature is not to blame for our misfortunes or our miscalculations, and since nature will not change in our lifetime, we must either accept the misery and cost brought about by weakened, bent spines, or we must do something about it.

- Redesign our footwear
- Exercise to strengthen and balance the muscles of our spines
- Modify our furniture, vehicles and workplace

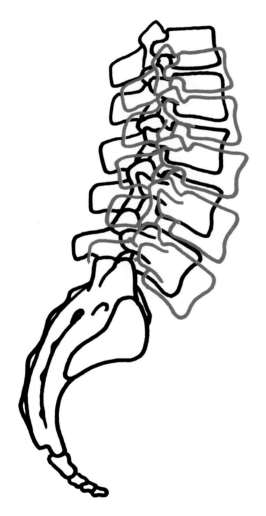

Figure 13-10. In the lordotic the lower lumbar vertebra tend to slide forwards and downwards.

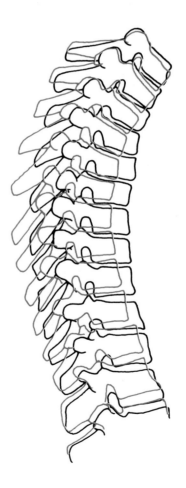

Figure 13-11. In the kyphotic the upper thoracic vertebra tend to be squeezed back.

> **Why does our poor posture cause so much back pain, so much misery?**

With some muscles of the torso being shortened, and others being weakened and stretched, the normal curves of the spine become exaggerated. The vertebrae tend to slide out of place, and the weight of the body increases this tendency. In the short sway back type, this occurs in the small of the back.

In the tall round shouldered type, it occurs at the top of the spine in the neck and upper chest region.

This distortion of our anatomy stretches or shortens ligaments, tendons, and capsules of joints; it may put pressure on nerves; it makes nipping and pressure on the disc extremely likely; and over the years it will produce the bony changes of osteoarthritis. Pressure on a nerve may produce pain, muscle weakness, pins and needles, lack of sensation in the skin, and headaches.

Where the main derangement is in the small of the back, the sciatic nerve, which runs down through the buttock to the foot is affected. The other main derangement, i.e. in the neck, will affect the nerves to the arms and head. The British royal family, inheriting a long and kyphotic neck, which is craned forward so frequently to listen and talk to people, is now in greater danger from osteoarthritis than from the executioner's axe.

How is a weakened and distorted spine going to affect you? You may feel tired towards the end of the day at the workbench or office, attributing this to mental fatigue or boredom, when in fact it is due to tired posture muscles. Your golf or tennis may not be as you would like, and you may not be able to garden and bend as you used to do. You may have experienced back pain or pain down your arm or leg. You may have had tablets from your doctor, which have given you relief. But these do not attack the cause, so sooner or later symptoms are likely to return, to increase, to become more constant.

The next treatment is likely to be manipulation. You might get instant relief. This may happen if one of the small joints of the spine has been partially dislocated, and it is put back into place. Or it may come from relaxation of muscle spasm, or tearing of ligaments to allow reduction in pressure. But, again, this is only relief, and symptoms will almost certainly return. Even if the manipulation is performed under an anaesthetic, and the spine can be forced back into correct alignment, it will soon revert to its old position under the pull of unbalanced muscles, and tightened ligaments and joint capsules. So that it is common to find people who have been manipulated repeatedly for years.

The next step may be traction, where your spine is forcibly stretched. Even if this could pull your spine into correct alignment, it would soon revert to its abnormal position, under the pull of muscle tension. You may then be placed in a plaster cast or brace to support your back. This is good treatment to give relief to a severe back injury. However, when a brace is used for a weakened and distorted spine, the muscles become weaker and weaker through disuse and the condition deteriorates.

Acupuncture which is sometimes recommended is only a pain reliever and can do little to correct the structural distortions which are causing the pain.

At this stage, surgery is likely to be offered. This usually entails joining two or more adjacent bones of the spine (vertebrae) rigidly together by a bone graft. Now, part of the spine is permanently and irreversibly fixed in a rigid position. The spine can no longer function correctly, and unfortunately, the muscular imbalance of the spine is now beyond correction.

This operation is intended to relieve pain or paralysis, which it may do. However, if it does not, the pain is liable to be constant. Finally, attempts may then be made to relieve the pain by deadening a nerve. If these fail, the pain is likely to be increased and permanent.

This is a terrible progression of misery, over many years. I do not believe that it is necessary.

During one 18 month period, I treated 68 patients with back problems by exercising the appropriate weakened muscles. Their ages varied from forty-four to seventy-nine, the average being sixty-one. This was an elderly group, and most had a long history of suffering, several for thirty years. In fact all but two, who were not x-rayed, showed x-ray evidence of osteoarthritis of the spine, confirming the long-standing nature of their disability. Several had suffered serious injuries to their spines. Most had already received different treatments, both by conventional and by non-medical methods. I did not include any who had

had their bones grafted together (fused), for I considered that muscle strengthening could not modify these cases. Twenty-two showed improvement with a reduction in their pain, while forty-three became symptom free. Of the three who failed to respond, one insisted upon doing exercises which counteracted mine, and two could not be persuaded that exercise would help. The conditions were not ideal for this treatment. Time for consultation was limited, as these patients were part of a very busy general practice, on an island where facilities were unsophisticated. With better conditions even more satisfactory results could, no doubt, be achieved.

Twenty-five years ago, when trying to treat blood pressure, angina, and those who had had heart attacks with exercise, I found the resistance of patients almost impossible to overcome. Now that this is accepted treatment the patient expects it. However, there was surprisingly little patient resistance to exercise for arthritis, even to the use of weights for the elderly.

The rapid relief of pain proved to be a great incentive. Patients would be even keener if this became accepted and conventional treatment.

Because no effective and permanent cure is offered by doctors to the legion of back sufferers, there can be little wonder that patients turn to treatments ouside conventional medicine. Yet, as these treatments do not tackle the cause of the condition, namely poor posture, it is not surprising that they usually only provide temporary relief, at best.

> **It should be noted that if any symptoms of back pain occur, the first step must be to obtain a proper diagnosis from a registered medical practitioner. Back pain may arise from any of a multitude of other diseases in other parts of the body, and the spine may well be perfectly healthy.**

One cannot talk about backs without referring to slipped discs, for this complaint has topped the poll for years. This is a misnomer as discs do not slip; they tear, or are cut, and then the soft inside bulges through the tear in the tough outer coat of the disc, putting pressure on adjacent nerves. The twenty-five discs which are fixed above and below to the bodies of the adjacent vertebrae, are cylinder shaped and made from a rubbery material, surrounded by an outer coat of tough fibre. They are fixed on to the bones so stongly that the bones themselves are likely to give before the outer casing of a disc would pull away from them. Disc injury probably occurs with a sudden rebound movement of the spine, usually performed as a defensive mechanism. When the spine is bent

Vertebra　　　　　　　　　*Tough outer coat of disc*

Figure 13-12. Two consecutive vertebrae with the intervening disc, when the spine is bent back as in lifting.

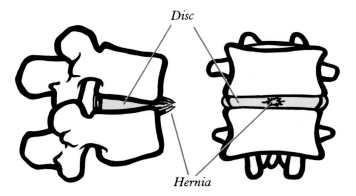

Disc

Hernia

Figure 13-13 and Figure 13-14. A sudden rebound of the spine may split the outer casing of the disc allowing the inner material to bulge through.

back, as in lifting, the disc alters shape to fit the new positions of the bones, and no disc injury will occur.

However, if the feet of the person who is lifting slip, or a load gets out of control, the lifter may bend forward suddenly and rapidly, in an attempt to regain balance. Although the rigid bones move instantly the pliable disc takes time to re-adjust its shape and a blow from the sharp edges of the bone may split the outer casing of the disc, allowing the soft inner material to bulge through.

No manipulation or acupuncture needle will repair this injury. With rest and the correct exercises, the body will make some repair, and there may even be a permanent cure, but surgery may often be necessary. Fortunately, a torn disc is not common, although it is frequently claimed and often misdiagnosed. Statistics on the misery and cost of back trouble are alarming. But, it is hard to feel statistics. It is better to look at the individual, if you wish to understand and sympathise.

In treating back problems, it has to be remembered that the back is part of the weight-bearing system, which also includes the hips, knees, ankles and feet. It is necessary to find out which joints are producing the misalignment. The knees may be bent because of weak thigh muscles, the hips may be bent due to continuous sitting, the spine may be bent because the person is to tall or too short, or the head may be twisted by poor vision, or deafness. Appropriate exercises must be prescribed to strengthen the weakened muscles. (Chapter 21).

CASE HISTORY

He is sixty-two. His first back pain occurred thirty-five years ago. For twenty-five years his back has been a major and constant problem, forcing him to make frequent changes of occupation and interfering with his whole way of life. He was given his first brace (corset) twenty-five years ago and has had two more. He spent three months in a plaster cast. He has lain on a board for months. He has had traction. For six weeks he lay on his back with weights attached to his feet to exert a constant pull on his spine. When he retired he faced a life of increasing pain and restricted movement. Within three weeks of treatment with exercise, he reported that his life had been changed. After nine months he reported his back to be "bloody marvellous after all these years". No longer an invalid he had converted a small seaside bungalow into a full house, concreted paths, grown vegetables, worked hard physically every day, and gone fishing.

Under such a regime there seems little reason why even the most long-standing and most intractable back problems cannot be helped. Whiplash injuries of the neck in car accidents come into this category and respond favourably. Claims that a back is wearing out are no longer tenable, for a back can be correctly aligned and strengthened at any age.

Many books dealing with back problems list dire warnings on *not* to lift or bend or twist. what a terrible indictment of 'Expert' advice! What a terrible way to live to have to be wary of every movement. Provide yourself with a strong supple back and you will be able to move in whatever way seems natural to you.

> **Backs can be treated simply, easily and at little cost. Most importantly, back trouble can generally be prevented.**

TREATMENT: See Chapter 21 for:
(A) **Essential Exercises 1, 2, 3.**
(B) **Recommended Exercises 5-10**
IN ADDITION:
(C) **If sway backed Exercises 22, 23**
(D) **If round shouldered Exercise 4.**
(E) **If true leg shortening — Raise whole shoe required amount**
(F) **If apparent shortening of leg Exercises 11, 12.**
(G) **If spine rotated Exercise 13.**

CHAPTER FOURTEEN
POSTURE IN OFFICE AND INDUSTRY

*"lo, this only have I found, that God
hath made man upright; but they
have sought out many inventions."*

Ecclesiastes chapter 7 verse 29

Neither governments nor industries, unions or employers have any real concept of the cost of muscle fatigue. The equation balancing machinery and equipment expenditure against productivity is regularly defined on the basis of unsound data, but labelled convincingly as ergonomics.

The human body is beautifully designed so that when standing or sitting its weight is borne almost entirely by the bony framework. This means muscle fatigue will occur only when muscles are being used for some other activity-provided, of course that the posture is perfect. Unfortunately, our present education system ensures that posture is ruined at school, and that the body has little chance to develop physically along natural evolutionary lines.

In office or industry the body must use equipment which makes posture even worse. Muscles are brought into constant play to counterbalance the undesirable positions into which the body is forced. Muscle fatigue quickly develops, resulting in inefficiency, accidents, discomfort and discontent. Until we design our machines to meet the postural needs of the worker, muscle fatigue, with all its dangers and deficiencies, will continue to occur.

The problem of worker efficiency is repeatedly being tackled, under the name of ergonomics, but, if the basic principles are wrong, ergonomics will not produce the desired results.

At a leading world institute of technology, with a major interest in ergonomics, the students are expected to work, study in the library, and sit examinations at flat-topped desks — light years behind the times when compared with the sloping desk. In the lecture theatre they have to look down at the lecturer instead of up. And, of course, heels on these students shoes make a mockery of ergonomic principles.

> **Workers with poor posture will fatigue easily no matter what production incentives or equipment are provided.**
> **If the worker's posture is perfect it will become distorted by badly designed equipment or work movements.**
> **Physical work cannot be performed safely and efficiently when heels are worn on footwear.**

Time should be allocated during working hours, to correct the posture of the worker. (Exercise sessions are already in force in some countries, and, although not specifically designed for posture, are an attempt to energise the worker and improve production.) For the first three months, this exercise programme should be specifically designed to correct posture only, and fifteen minutes a day does not seem too high a price to pay for greater comfort, reduced fatigue and injury, and greater production (Chapter 21). After three months, when posture should be good, the programme could be switched to exercise for cardiovascular fitness and general muscle tone. Such sessions counteract the harm to the worker of our modern industrial life, and promote the worker's well-being and health. The young, on entering the workforce, ought to have good posture. Pressure from unions, management and parents on the controllers of the education system would help to achieve this goal.

In addition to correcting bad posture and providing sufficient muscular power and reserve to do the work, there are two main principles which govern work design:
1. The head must be aligned correctly with the spine.
2. Where possible, muscle use on one side of the body should be balanced with muscle use on the other side.

THE OFFICE

It is little wonder that office work causes a tendency to develop a dowager hump, arthritis of the neck, a curved spine, and arthritic fingers.

Chair and desk design have already been discussed in the chapter on schools. For chairs the vital factor is to provide support for the whole length of the thigh. So the seat has to be adjustable in length and it also needs to be tilted backwards slightly.

The desk should have a writing surface which can be adjustable almost to the vertical position. A magnetised metal strip down one side would hold pens, rulers, protractors and other small instruments, provided that they are made containing a small piece of metal. There are many alternatives for holding paper in position on the incline surface; clips on the top or on the sides; rubber bands across

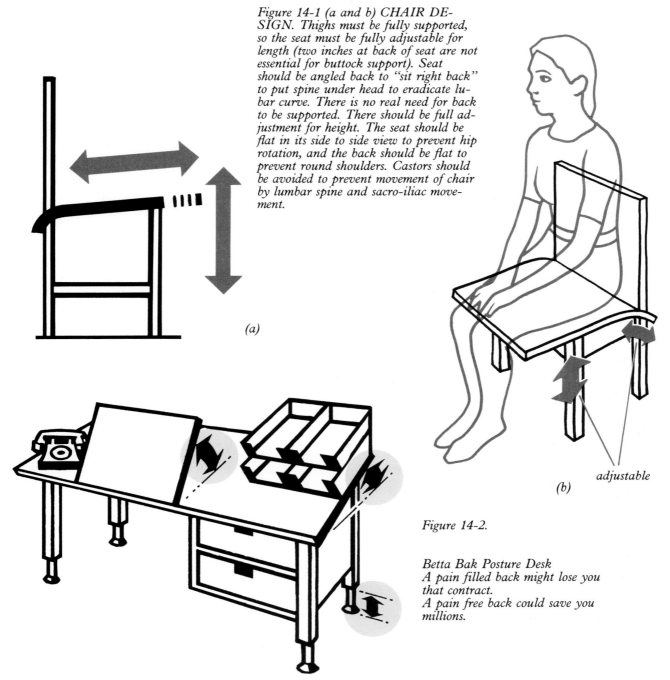

Figure 14-1 (a and b) CHAIR DE-SIGN. Thighs must be fully supported, so the seat must be fully adjustable for length (two inches at back of seat are not essential for buttock support). Seat should be angled back to "sit right back" to put spine under head to eradicate lumbar curve. There is no real need for back to be supported. There should be full adjustment for height. The seat should be flat in its side to side view to prevent hip rotation, and the back should be flat to prevent round shoulders. Castors should be avoided to prevent movement of chair by lumbar spine and sacro-iliac movement.

(a)

adjustable

(b)

Figure 14-2.

Betta Bak Posture Desk
A pain filled back might lose you that contract.
A pain free back could save you millions.

the surface; or the paper could be fed through rollers at the top or bottom or at either side. The remainder of the desk should be at an angle of about thirty degrees to the horizontal. If the surface is made of a non-slip material, papers, files and books will remain in place and will be much more easily visible and reachable.

On one occasion I was going through the design office of an international company. The designers worked at flat desks, illuminated from below, for their work was very detailed and required extreme accuracy. My suggestion that they tilt their desks to the nearly vertical position was not well received. However, in one corner was a designer working with his desk tilted in this way. He explained that he had lost eighteen months off work through recurrent back problems, until, finally his family doctor had suggested that he tilt his desk. After following this advice his back had been uncomplaining.

Until such desks are freely available, it is easy to improvise your own. A piece of board, a couple of rubber bands, some old telephone books as props, and a piece of rubber on the desk top to stop the board from slipping, and you will have a writing surface to ease your back, and to enable you to sit with your back and head straight while writing. You will soon be aware of the absence of muscle fatigue and eye fatigue at the end of the day.

The typewriter should be modified to reduce the amount of fatigue and posture distortion it causes. The surface of the keys themselves should not be horizontal, but angled towards the operator. In fact, the whole machine needs to be angled forwards, and this can be achieved by raising the back 10-15cms.

Figure 14-3. Tilting a typewriter (either by the designer or by propping it up on books) and a copy holder above the typewriter will improve posture and reduce fatigue.

The roller for the paper should be higher, almost at eye level. There should also be a stand above the typewriter and at eye level, on which the copy can be held. Stands like this are available at small cost.

In typing the fingers are held almost rigidly with the end joint of the fingers in a slightly bent position. In time, these finger joints lose their ability to flex or extend, and are held in a slightly bent rigid position, so that osteoarthritis develops in these joints. Finger exercises will prevent or correct this. (Chapter 21)

Much other office equipment needs redesigning, using the principle that the operator should be able to work the machine with the head held in a direct line above the spine, so that all the weight is being borne by the spine. If twisting to one side cannot be avoided, try to arrange the work so that the sides can be alternated. Work that must be done standing should be carried out at eye level with the working surface tilted towards the worker.

The investment, the brains and the competition in the computer industry is enormous. Yet, little thought appears to have been given to the posture, the comfort and, therefore, the efficiency of the operator. The keyboard should be tilted towards the operator, so that the keys can be seen and struck more easily without hanging the head forward. The screen should be just above eye level (adjustable). The head will then be held over the spine, and not bent forwards, and the eyes will look up. (Chapter 16 explains the deterioration of eyesight caused by office work.) The high screen will help to prevent this deterioration, and reduce everyday eyestrain.

INDUSTRY

If the human body were as inflexible as the man-made machine, we would encounter few problems of back strain, muscle fatigue, or arthritis, for it would be impossible for a rigid spine to mould to a machine, as our flexible spines do. We would therefore have to design machines to suit our anatomy, rather than the other way round, as in fact happens. Because our bodies can take on the shape required by our machines, designers seldom bother to fit machines to our needs. We are expected to mould to the machine — not because we should, but because we can, and this makes it easier for the machine designers who sometimes falsely claim that their machines are ergonomically correct.

Our bodies are not designed to carry out an awkward manoeuvre, over and over again, all day, every day. Yet this is what industry frequently asks its employees to do. Where it is necessary for an operator to repeat continuously a movement that throws the spine out of alignment, modifications are necessary to improve efficiency, decrease accidents, and keep the operator happy. Consider a seated worker, who has to bend forward and pull a lever on the left, and at the same time, crane the neck forward in order to see a gauge. Doing this movement once every half hour is very good exercise, but doing it every two minutes will cause back and neck troubles, and eventually arthritis will develop in the left hand, wrist, elbow, shoulder, neck and back. In the first week the worker is likely to suffer from inflammation of the tendons of the muscles that are being used. After about a week, the muscles will respond to the stimulus with increased efficiency, and the operator will be able to do the work without apparent harm. But all the joints will be tending to bend toward the line of activity, so bending to the right (opposite to the line of greatest use) or straightening up will begin to cause trouble. And towards the end of the day muscle fatigue will start to occur in the over used muscles.

> **The time to avoid these problems is at the design stage of the machine or procedure.**

The designer should try to visualise an operator with a spine held rigid in its correct curves and with the eyes looking straight ahead. This is the basic design requirement for any repetitive work, along with the need to balance the work on both sides of the body. Then he will have the operator sitting or standing with the head correctly aligned above the spine. If we redesign our original machine, there will be a lever on either side which can be operated by either hand or both hands together. Levers should also be operable without bending forward. This means that either the seat or the lever height should be adjustable. Gauges and dials should be placed directly in front and at eye

level. Many machines in use today could be satisfactorily modified for operator comfort.

INDUSTRIAL FOOTWEAR

A major contribution to worker comfort, efficiency and safety would be the provision of working footwear without heels. The economics are clear. The cost to industry of back trouble and unnecessary fatigue induced by the wearing of incorrect footwear far outweighs the cost of providing correctly designed footwear.

The problem is to persuade its use. Many forms of protective clothing are now accepted as compulsory by union and management. The welder's face guard and the hair net of the machine operator are the norm. In time, heel-less footwear could become routine.

Seated workers will argue that they do not need such shoes, but they still face the prospect of spinal distortion during the period that they are on their feet. Management might think that it is not worthwhile providing heel-less shoes for their workers unless they are worn in leisure hours, as well. They need not fear, for if heel-less shoes are worn eight hours a day for a month, heels will make people feel so uncomfortable and ungainly, that they will be reserved for use only on special occasions.

In standing occupations correct shoes are vital, especially where fitness is needed, such as in the Police and the Armed Forces. Viewed from close up, the soldiers of the Guards Regiments who take pride in their postural perfection have obvious deficiencies of posture, as shown by the gaping pleats in the upper parts of their great coats, and the hang of those coats. The combination of their height and heels produces a tendency to kyphosis (Chapter 5). In an effort to obtain perfect posture they are provided with off duty caps in which the peak points vertically downwards with the aim of making them hold their heads up. This actually, makes them tilt their heads backwards out of alignment.

TRANSPORTATION

One of the major problems in the transport industry is that continual sitting with the hips bent produces weak back and buttock muscles, and this is aggravated by working the foot pedal. This weakness used to be counteracted by the activity of loading and unloading the vehicle, but in industrialised countries this is often performed by machines, tail board lifts, people from other unions, or the use of containers.

Some years ago a truck driver collapsed at the wheel with acute heart failure, in the middle of a tunnel under the Thames in the rush hour, and caused a massive traffic jam. On recovery, he explained that he had gained sixty pounds in weight rapidly, since the nationalisation of the British Transport Industry. The significance of nationalisation was not apparent, until he explained that he was no longer permitted to unload or load his twenty ton truck.

Where industry produces dangerous fumes, masks are provided and expected to be worn by the worker for his own health and safety.

> **Where industry, or the office, prevents natural daily physical activity, which is vital to our health and posture, it should be mandatory for time and facilities to be provided for such activity.**

Just as the worker must accept the wearing of a mask when working in dangerous fumes, he should accept the need for physical activity, when in a sedentary occupation.

The upright and firm seat of the truck is reasonably good design, so that fatigue is far less than in the low, softer inclining seats favoured by car designers. However, in both cases it is essential that the weight of the thighs be supported fully, to prevent fatigue, and distribute the weight over a wider area. This is especially important for the leg that works the accelerator. To make sure of this support, the seat, rather than the back should be adjustable for both angle and thigh length as well as height.

The insistence on good lumbar support, i.e. support in the small of the back region, whether in a car seat, or any other seat is illogical, and has been discussed in Chapter 8 (see Figure 8-3).

Attempts to "keep the back straight" when sitting are mechanically unsound, which is why the deck chair has always been so comfortable, and the

full support this chair gives to the thighs eradicates the need to cross and uncross the legs.

The recommended position for holding the steering wheel at ten o'clock and two o'clock, is illogical for both postural and safety reasons. It makes a person round-shouldered, and by forcing him to raise shoulders and arms, produces fatigue. In an emergency, it is unstable, as the driver tends to be hanging on to the wheel rather than controlling it. If the wheel is held at seven o'clock and five o'clock, the shoulders can be dropped and the upper arms can hang vertically, reducing fatigue.

Figure 14-4. The ten o'clock, two o'clock hand position is unstable and tiring.

Figure 14-5. The seven o'clock, five o'clock is stable and relaxing.

In an emergency, the position is stabilised by the shoulder and elbow against the door, and the grip is very strong.

Why should we all be forced to buy cars with a low, racing, stream-lined appearance and low uncomfortable seats, guaranteed to produce aches and fatigue, especially in the tall or short individual? As in so many manufactured goods, we are offered what the designers think is marketable and then advertising is used to convince us that we really want it. Some of us want a car that is easy to get in and out of, that is a pleasure to drive, that has excellent visibility, that will not produce a sore back and aching joints and will not fatigue us.

We have the choice of a London taxi, a light truck or a vintage car. Manufacturers will tell us that, if the car is not streamlined it will have less acceleration, lower power-weight ratio, and a greater fuel consumption! However, any extra fuel cost is likely to be balanced by reduced doctors bills, accidents and insurance premiums. Streamlining for trucks and buses would seem to be more logical, but no truck driver would put up with such uncomfortable seats. Ironically, when delivering this manuscript to a London publisher, the taxi driver, unaware of the contents of my parcel, told me how he could drive his taxi all day without fatigue, but that an hour in his own car caused backache.

The fatigue of international air travel is notorious, and for the business man or statesman it may be disastrous. It is produced by such things as time change, immobility, boredom, pressurisation and muscle fatigue from poor seat design. To reduce muscle fatigue it is necessary for as much of our body weight as possible to be supported by our bones, rather than our muscles. The seat needs to be angled backwards slightly, so that the buttocks are right back in the seat, and underneath the weight of the body. The whole length of the thigh should be supported by the seat, so that the weight of the body is distributed over the buttocks and the length of the thighs. People with long legs can never find a chair with adequate support to the thighs, which is why they cross their legs, when at least the thigh of the upper leg is gaining full support. So the seat needs to be adjustable for length and tilt. The head-rest should be adjustable for height and rake, so that the head can rest in its natural position.

Most travellers spend a lot of their time reading, with the head bent forward to look down at the book. This throws the whole seating posture out of line, and is a major cause of muscular fatigue. The usual folding table could be easily adapted to double as a book rest at eye level. Then the passenger could hold the head upright, with the spine in alignment, throughout the hours of reading.

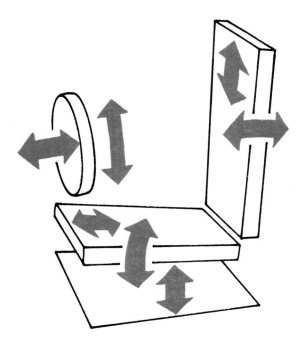

Figure 14-6. Some of us would prefer ease of exit and entry and freedom from backache and fatigue to speed and appearance — if given the choice by manufacturers.
Seat is adjustable for height, depth and rake. The back for height. The steering wheel for height and rake.

Exercises should be advised for long haul passengers. To counteract the muscle fatigue of sitting, to loosen joints maintained in one position for many hours, and to stimulate circulation.

Cabin staff experience frequent back problems causing much pain to the sufferers and considerable cost to the airlines. Much of this could be prevented by removing the heels from their shoes, and by corrective and preventive exercises — which would improve their grace and posture.

When the principles of good posture are understood, a great deal will be done by industry and commerce at a very small cost relative to the economic and physical benefits brought about by these changes.

Figure 14-8. Aircraft seat. Comfort for the long-legged, at last. Thighs supported by adjusting the depth and tilt of seat. Book rest to counteract fatigue.

Figure 14-7. Aircraft seat. Discomfort for the long-legged. Knees jammed, thighs unsupported.

TREATMENT — See Chapter 21 for:
(A) Essential Exercises 1, 2, 3, 4
(B) Recommended Exercises 5, 6, 7, 8, 9, 10
(C) Fingers Exercises 17 and/or 18.
(D) Wrists Exercise 16.
(E) Elbows Exercise 14.
(F) Shoulders Exercise 15.

CHAPTER FIFTEEN

POSTURE IN SEXUAL INTERCOURSE

*"Is it not strange that desire should so many years
outlive performance"*

William Shakespeare (1564-1616)

In a purely physical sense sexual capacity, performance and satisfaction is governed by two factors — posture and circulation.

As sexual intercourse has been refined by nature over millions of years to ensure that the species survives, we can be sure that the closer sexual performance keeps to nature's design, the more pleasurable it will be for both partners.

Infertility is becoming a major and accepted problem. Research on this problem is following the expected line of searching for a psychological or hormonal answer, when it lies in low fitness levels producing inadequate and infrequent desire coupled with a weak erection, plus poor posture producing inadequate penetration. The end result is less mutual satisfaction which in turn leads to a reduction in the frequency of sexual intercourse.

For those wanting an abortion, becoming pregnant seems only too easy, but for the childless it appears an impossible task. In truth it is a fantastic voyage! There are about 120,000,000 sperm to every cubic millilitre of semen, so imagine how small they are! And they have to travel a distance of somewhere between eight and eleven centimetres to find and fertilise the egg. It is about as chancy as swimming the Atlantic and arriving in New York in time for breakfast. Nature is very economical. That is how she maintains the balance of the species (the ecology of life). Yet, she has to produce thousands of millions of sperm, propelled with a force which would send them several feet through the air, in the hope that one will reach the egg, for only one is required. And she makes sexual desire frequent for the physically fit, so that repeated attempts at fertilisation are made — thus increasing the success rate.

To increase the probability of pregnancy, nature's evolutionary aim is for the tip of the penis to be as close to the entrance of the womb as possible, at the moment of ejaculation, and on these principles whatever method produces this will be the most enjoyable for BOTH partners.

I believe that the penis can enter the neck of the womb (cervix) during intercourse especially when good posture produces deep penetration. It is considered that sperm reach the egg, in the tube, faster than they can swim there, and it is suggested that contractions of the uterus may assist their movement. However, if the penis has entered the cervix at the time of ejaculation, the speed of travel of the sperm is no longer a problem. The force

with which sperm are ejected seems quite unnecessary, if they are only meant to be deposited in the vagina.

POSITION AND POSTURE

We are unique among the animal world, for we have a choice. For four-legged animals, there is only one method — standing rear entry. Even monkeys, who are capable of standing on two legs (but not upright) and walking on two legs, are incapable of any face-to-face position. For their thighs are so bent forward at the hip joint that they cannot get close enough to each other for insertion of the penis. There is so little forward curvature of the spine in the buttock region, that the vagina is pointed backwards and away from the penis when in a face-to-face position.

Figure 15-1. Monkeys cannot straighten their upright posture sufficiently for face to face sexual intercourse to be possible.

Humans are the opposite. It is quite impossible for us to have rear entry sex standing upright, for the entrance to the vagina is on the other side of the body to the penis. However, humans can bend at the hip joint and put themselves in the four-legged posture of animals, bringing the vagina back to face the penis, so that rear entry sex can be performed.

Humans have a marked curve at the lower end of the spine, so that the vagina is to the front of the body and in rear entry sexual intercourse no part of the male body comes in contact with the female's most sexually sensitive region, the clitoris. Also very little pressure can be exerted by the male body on the vaginal lips, which are also highly sensitive.

It has been implied that a preference for the rear entry position indicates the tendency to revert to the position of animal ancestors. This is nonsense, and shows a lack of understanding of the process of evolution. We have shown no preference for eating on four legs, or opening our bowels or bladder in this position, so why pick on sex? Bent and inflexible hips, and the interference of fat bellies, are making the rear entry position more popular — not any deep psychological regressions or atavistic tendencies.

The front wall of the vagina is shorter than the back wall, and the top of the erect penis is shorter than the underside. In front entry sexual intercourse the shorter lengths are in contact with each other. Entry is therefore deeper and contact between penis and womb is closer than in rear entry.

As the entrance to the womb is tilted forward on the vagina, and the urethra (the tube through the penis) is closer to the under-surface of the penis than the top surface, ejaculate is more liable to enter the womb when using a front entry position. With rear entry it is more likely to impinge on the front surface of the cervix (the neck of the womb).

At the moments of orgasm, the female thrusts her pelvis violently forward. In a front entry position this pushes the penis deeper into the vagina, but tends to pull the vagina away from the penis with rear entry.

> **Our anatomy and our posture indicate that humans are more likely to conceive with front than rear entry sex.**

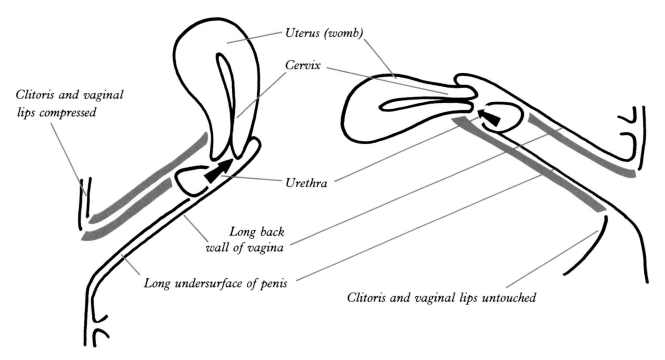

Figure 15-2. *In front entry sexual intercourse, the shorter upper surface of the penis is in contact with the short front surface of the vagina. Penetration is deeper. Pressure on the clitoris and vaginal lips takes place, and ejaculate is more likely to enter the womb. (Red denotes shorter wall of vagina, and shorter surface of penis.)*

Figure 15-3. *In rear entry sexual intercourse, the longer surface of the penis is in contact with the shorter wall of the vagina. Penetration is not so deep, entry of semen into the womb is less likely and pressure on the clitoris and vaginal lips does not occur. (Red denotes shorter wall of vagina, and shorter surface of penis.)*

The sensitive parts that a woman likes to have stimulated are first and foremost the clitoris, followed by the front two thirds of the inner lips of the vagina, the entrance to the vagina, the perineum (the area between the vagina and the anus), and to a lesser extent the outer lips.

For the man they are the tip of the penis around the hole (the urethra), the bulbous end (the glans), the underside particularly near the end (the frenulum), the base of the penis, the front of the scrotum (the sack containing the testicles), and the top of the shaft of the penis (the Corpora Cavernosa).

The sexual organs of the two sexes are composed of identical structures. The difference is only in relative size, and the fact that the female equivalent of the underneath of the penis (the Corpora Spongiosum) has been split in half to permit the presence of the vagina.

It is not surprising, therefore, that the areas of sexual arousal are the same, with an identical pattern of increasing sensitivities, except that the equivalent of the perineum of the female is the skin of the scrotum. It would be surprising, therefore, that if man obtains his maximum satisfaction during insertion into the vagina, that woman does not achieve her maximum under the same circumstances.

With the common front-to-front position these areas are mutually stimulated, (provided the male has good posture) and each partner can vary the amount of stimulation obtained, by the closeness and pressure exerted. In particular the woman's clitoris is pressed against his pubic bone just above the penis. There is a depression (called the vestibule) between the clitoris and the vagina, into which the male pubic bone slots, enabling the woman to obtain more pressure on the underside of her clitoris. In fact, the clitoris becomes squeezed and compressed between her pubic bone and his. So, like the penis it is being stimulated from above and beneath. The outer and inner lips of the vagina press on the male pubic bone on either side of the penis, and this pressure is supported by the female pubic bone behind these lips. The perineum of the female comes in contact with the front of the scrotum and as there is no bone supporting either of these structures, the sensitive testicles are not hurt or damaged. The swollen and congested vaginal entrance and lips grip the base of the penis and the shaft during movement. Thus, all

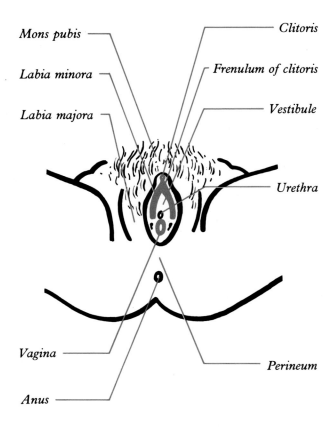

Figure 15-4. The sensitive parts of a female. The most sensitive parts are in red.

the vitally sensitive regions of both partners are stimulated simultaneously, as one would expect evolution to arrange.

Furthermore, the bodies can be held in close contact, breasts can be stimulated and kissing can take place.

In the early stages of sexual contact, the vital areas are sensitive to light touch. But as arousal

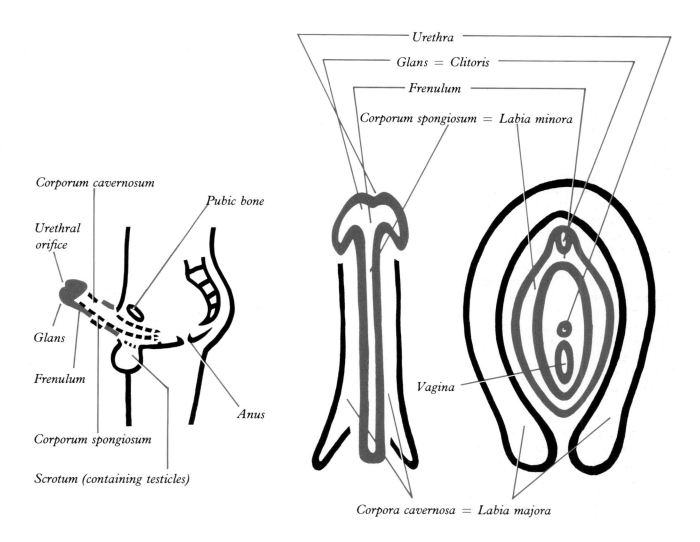

Figure 15-5. The sensitive parts of a male. The most sensitive parts are in red.

Figure 15-6. Structures forming penis.
Figure 15-7. Structures forming entrance to vagina.

proceeds, nature has ensured that this sensitivity is reduced, and is replaced by an urge for increasing pressure. Thus the thrusting movements of both hips increase, the female frequently gripping the male's buttocks with her hands, or crossing her legs behind him, in an attempt to get further on to him, ensuring that the penis is further inside her, and that the areas of stimulation are closer together.

In the final convulsive, uncontrollable thrusts of orgasm, both hips are rammed forwards to meet each other and produce maximum sensation. Thus the whole mechanism is designed to bring the penis and the womb as close together (usually touching) at the point of ejaculation, to increase the likelihood of pregnancy. Nature offers the biggest incentive of all. Sexual satisfaction in return for

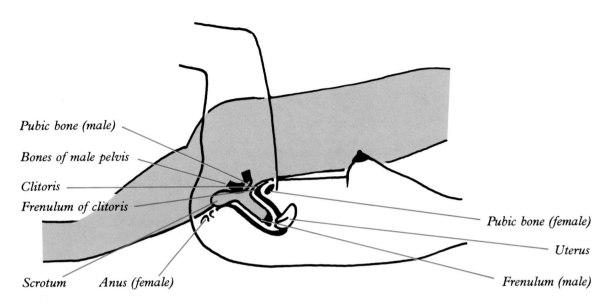

Pubic bone (male)

Bones of male pelvis

Clitoris

Frenulum of clitoris

Pubic bone (female)

Uterus

Scrotum   Anus (female)

Frenulum (male)

Figure 15-8. Face to face sexual intercourse, when the male posture is good, producing good contact and compression of sexually sensitive parts. Female parts in red.

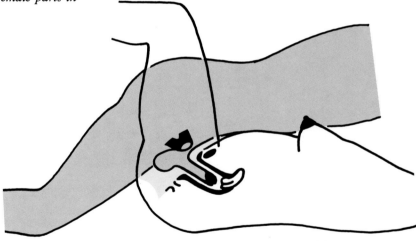

Figure 15-9. Face to face sexual intercourse, when posture is poor. Note the lack of contact and compression of the clitoris and inner vaginal lips of the female, and the base of the penis in the male.

pregnancy. However, if posture is poor the mechanism will not work properly.

When the male has poor posture, his thighs become bent forwards at the hip joint and he is unable to straighten this joint. The penis is pulled back from the depths of the vagina and full pressure of the male pubic bone on the clitoris cannot be achieved. The worse the posture, the more this process occurs, until the clitoris is not touched at all. A fat belly, which is likely to accompany poor

posture, makes the problem worse. Extremely bad posture necessitates regression to the posture of the apes, with front entry becoming almost impossible.

There is a modification of bone design between male and female which facilitates the close approximation of the sexual organs. We know that the first piece of the thigh bone (the neck) goes out sideways and horizontally from the hip joint of the female, but is angled backwards and downwards in

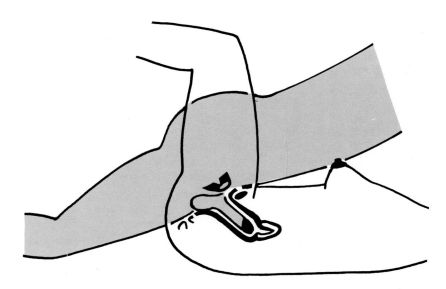

Figure 15-11. Face to face sexual intercourse, when the male has extreme hip flexibility. It is suggested that the cervix may open to allow entry of the penis, increasing the probability of conception.

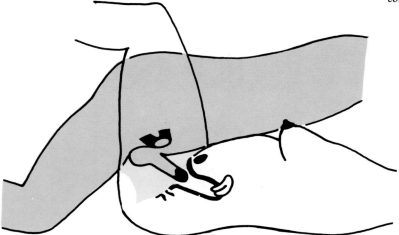

Figure 15-10. Face to face sexual intercourse becoming impossible, because of bad posture.

the male. This makes the woman's hips wider and the legs further apart, (in fact there is a gap between the top of the thighs), and makes the male hips slimmer, so close contact is easier. In bent, bad posture the man's hips are widened reducing close contact.

The advantage for sexual satisfaction of this skeletal variation is mechanically obvious.

The posture of the female is not so important, for bending at the hip joints assists insertion. But, with poor male posture, the unfortunate female is unsatisfied and frustrated. (See Figures 15-9 and 15-10.) Conversely, if the male can hyperextend (overstraighten) his hip joints, pressure becomes maximum and mutual satisfaction enhanced.

Investigators, having found that women are being inadequately stimulated, are advising in their manuals that the clitoris should be stimulated before, during, or after insertion, using hand or mouth. While there is nothing wrong with these actions the "experts" have fallen into the same trap as the advisers on backs, who blame evolution for our back problems, and advise an abnormal solution to the problem. Because our poor posture prevents stimulation of the clitoris by the male pubic bone, as nature has designed it, they infer that the design is inadequate and advise the separate stimulation of the clitoris. In fact, the harmonious and mutual actions of sexual intercourse, which should result in a satisfying and simultaneous orgasm for both partners and optimize the possibility of conception, can only be fully realised when postural perfection allows for deep penetration and maximum clitoral stimulation.

POSTURAL RELATED SEXUAL PROBLEMS:

If people are racked by the pain and immobility of osteoarthritis, the joy of sexual intercourse can be greatly diminished or even totally unobtainable. It has been suggested that the major cause of the cessation of sexual activity is osteoarthritis, rather than the ageing process itself. Back pain may either prevent sex or be produced by it. The sight of bodies sagging because of poor posture and poor physical condition is certainly a sexual turnoff. Postural obesity bringing two fat bellies together can be very frustrating. In fact, excessive fat in the genital area of the male makes the penis effectively much smaller than it really is!

SEX AND CIRCULATION

Without adequate circulation, sexual intercourse is difficult, unsatisfactory and in many cases impossible. The erection of the penis is brought about solely by pumping blood into the veins of the body of the penis. The swelling of the lips, breasts and nipples, and the engorgement of the vagina and clitoris are all brought about by increased blood flow and all add to stimulation and sexual satisfaction.

I have been amazed at the number of young men who have come to me in recent years complaining of impotence. There seems no doubt that this condition is more common and occurs at a younger age than before, even allowing for the fact that the young are less reticent than they used to be. It is senseless to pump them full of hormones when a simple solution exists. Many of these men (often in their twenties) are found to be suffering from heart and circulatory problems. If they cannot maintain an adequate supply of blood to keep their hearts going, how on earth can they expect to raise and keep an erection? One 'HiTech' solution to the problem is to insert a hand operated pump into the penis, to produce an artificial erection. This "bionic penis" must be the very height of spare parts surgery gone mad!

> **The effective answer to this particular problem, at any age, is simple. Regular physical exercise of sufficient severity to adequately stimulate the circulation will not only satisfy the obvious requirement for blood to the erectile tissues but also stimulate adequate hormonal production.**

Since the jogging explosion repeated letters to running journals tell of enhanced sexual performance.

The young now have the time and the money to drink more heavily, and it is not uncommon for teenagers to require treatment for alcoholism. Shakespeare described the problem perfectly when he wrote that alcohol increases the desire, but reduces the performance. Another problem affecting circulation is that of drug addiction. Addicts of all types invariably have reduced sexual desires and ability. Even nicotine has a marked effect on the circulation.

On the whole, the young are sexually unfit because their circulation is so poor. It is not surprising that they need more and more sexual stimulation to produce arousal.

Having developed the basic essentials of perfect posture and good circulation we can confidently exploit the subtleties of sexual variation without fear of failure.

Evolution has given us our hands which we can use for stimulation, our upright posture which permits a variety of positions, and our developed brain and imagination which we can use to the utmost. Because "meat and two veg" will keep us alive and well, there is no reason why we should not try and enjoy gourmet food. So use your imagination and the advice in the sex manuals for variations and unusual delights. However, if we dine constantly in restaurants we will soon wish for home cooking. In the face to face position nature provides for the maximum simultaneous mutual enjoyment and neither partner feels left out even for one moment.

> **But, if sex is to be as mutually satisfying as nature intended, the woman should demand that the man pay as much attention to his posture as to her clitoris.**

TREATMENT: See Chapter 21 for:
(A) Essential Exercises 1, 2, 3.
(B) Recommended Exercises 5-10.
IN ADDITION
(C) For Swaybacked Exercises 22, 23.
(D) For circulation — stamina training.
   e.g. jogging.

CHAPTER SIXTEEN

POSTURE IN EYESIGHT

"Why has not man a microscopic eye?
For this plain reason, man is not a fly."

Alexander Pope (1688-1744)

Of all the artificial aids to the mechanism of the body, of all the props and crutches, of all the spare parts, none is as common as the pair of spectacles. They are an accepted part of modern society to the extent that it is not even realised that the body's natural mechanism is being manipulated. Fortunately, no surgery is required, and the wearer can remove them at will.

Eyesight can be affected by heredity, by disease, by injury. Posture plays little part in such cases, but posture is responsible for much of the deterioration of eyesight which occurs with age.

There are four ways in which modern living has changed the way we use our eyes, compared with the use evolution designed.
● We do an excessive amount of close work — writing, reading, typing and working at machinery — with very little looking into the distance. So many city dwellers have little need to use their distance sight.
● We rarely use the special part of the eye that is designed to see with the greatest accuracy, because we rarely look for minute detail.
● We spend long periods looking in one direction. without turning the eyes to either side.
● We spend most of our time looking down, because almost nothing is designed with correct posture in mind.

When light rays reach the eye, they are bent first of all by the cornea, and then by the lens, so that they meet on the retina, the seeing part of the eye.

If light rays do not meet on the retina, there is blurred, indistinct vision. In order to ensure that these rays terminate precisely on the retina, a process called 'accommodation' takes place. This is

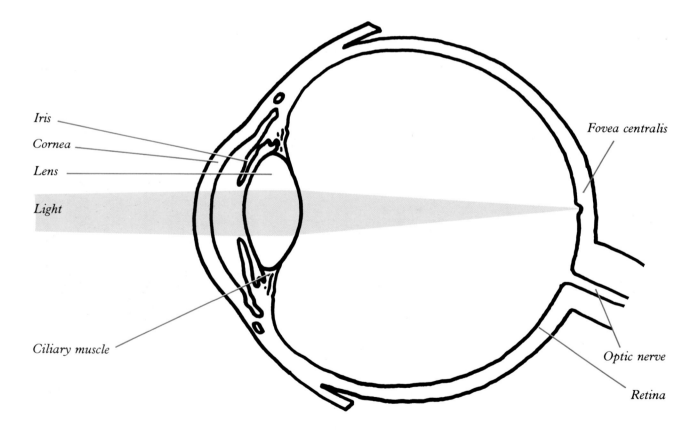

Figure 16-1. The eye, showing how light rays are bent by the cornea and the lens to focus on the retina at the back of the eye.

normally achieved by the internal muscle of the eye (the ciliary muscle) changing the shape of the lens.

Accommodation could also be achieved by moving the lens and the retina closer to each other or further apart. This is the method used by many animals and fishes, the change in length of the eye occurring as a result of the contractions of the external muscles of the eye, which move the eye in different directions.

The generally accepted opinion is held that all accommodation in humans is performed by the lens changing shape. But because the coats of the eye are non-rigid, contraction by either the internal or external muscles of the eye must change, to some extent the length of the eye. In fact, experiment has shown that when the circular ciliary muscle contracts the eye lengthens. So a small part of

accommodation, even in humans, must be achieved by change in eye length.

> **With the muscular imbalance in eye muscles, which develops owing to our way of life and posture, accommodation can be seriously affected!**

The curve of the cornea remains constant so it plays no part in the 'accommodation' process.

Sometimes the lens may become opaque (cartaract) and may be surgically removed. Nevertheless, accommodation may still take place, so this then *must* occur by changing the length of the eyeball.

The lens is surrounded by a circular muscle

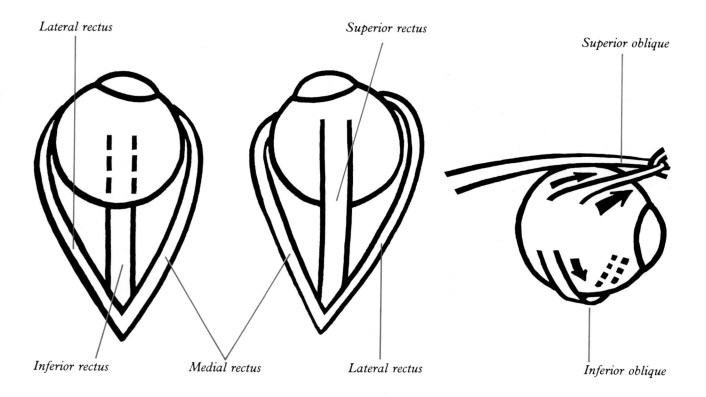

Lateral rectus

Superior rectus

Superior oblique

Inferior rectus   *Medial rectus*

Lateral rectus

Inferior oblique

Figure 16-2. The rectus muscles, four of the muscles which move the eye.

Figure 16-3. The oblique muscles, the other two muscles which move the eye.

(the ciliary muscle) which on contraction and relaxation changes the shape of the lens. There are six external eye muscles. Each is attached to the bony eye socket at one end and to the outer surface of the eye at the other. They move the eye in different directions, and, since the coats surrounding the eye are of non-rigid material, contraction of these muscles must change the shape and length of the eye. If nature intended that the eye remain constant in shape, then we could expect its outer coat (sclera) to be made of rigid material such as bone. But it is not. In fact, the outer coat contains elastic material — a definite sign that it changes shape.

It is generally contended that the elasticity of the sclera (outer coat of the eyeball) is required to keep the pressure inside the eye correctly adjusted. This may well be its main purpose, but the fact that this elastic material is there ensures that the eyeball shape and length must change, when either the circular muscle (ciliary muscle) or the external muscles contract. And, further, that the shape will return to its neutral shape when these muscles relax, for these are the properties of elastic material.

The combination of movement which can be obtained by the six external eye muscles acting on each eye is very complicated indeed. Four (the recti) are attached in front to the front half of the eye, and at the back to the back of the socket, — although offset to one side, i.e. towards the nose.

When they contract they pull the eye further into the socket and shorten its length.

The other two muscles (the obliques) are attached to the back half of the eyeball and pull towards the front of the eye socket.

While it would seem obvious that the length of the eyeball can be varied by contractions of the external eye muscles, and by the elasticity in the sclera, exactly which combinations of muscles are best for shortening and for lengthening the eye have yet to be determined.

If we spend excessive time looking down and little in looking up, the down-turning muscles (superior oblique and inferior rectus) become well developed and shortened, whereas the up-turning muscles (inferior oblique and superior rectus) become weak, lengthened and tend to waste (degenerate).

We require the combined action of these muscles to lengthen or shorten the eye. Because we spend so much of our time looking down, muscular

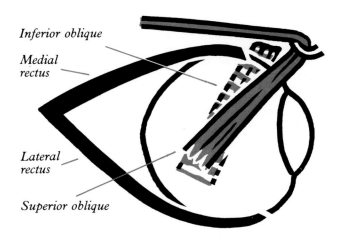

(a)

Pull of side muscles squeezing eye — Medial and lateral rectii.

Inferior oblique

Medial rectus

Lateral rectus

Superior oblique

(b)

Pull of upper and lower muscles squeezing eye — Superior and inferior recti and superior and inferior oblique.

Superior oblique

Superior rectus

Medial rectus

Inferior rectus

Lateral rectus

Inferior oblique

Figure 16-4. The right eye seen from above (a) and the left eye seen from the side (b). The obliques try to pull the eye out of its socket. The recti try to pull it into the socket.

imbalance develops and increases with age. Many people cannot then bring near objects into focus and need reading glasses. As the muscle imbalance progresses, even distant objects cannot be focused, and bifocals may be required. This muscular imbalance reduces not only the ability of the eye to lengthen and shorten, but also reduces the amount of elastic material in the outer coat (the sclera) because of reduced activity, i.e. reduced stimulus. The imbalance further ensures that the ciliary muscle is fixed more rigidly in its lower half than in its upper half, so that changes in the shape of the lens will be uneven.

Experiments have shown that if an animal has only a rudimentary inferior oblique muscle the eye cannot lengthen, and when we weaken ours, by looking down, we tend to suffer in the same way.

It is our changed way of life that has caused us to look down too much. The normal way of walking is to place the heel on the ground first and then roll onto the outside of the foot, then the sole of the foot, and finally onto the big toe. With heels on shoes, this is very hard to do properly and there is a possibility of tripping. The rigidity of shoes means we have to walk on smooth surfaces or else be very careful where we put our feet because the shoe does not mould easily to an uneven surface. We therefore tend to look down when we are walking, and watch where we step. In fact, most of our day is spent in looking down. We are generally forced to do so by our work and pastimes. Our television is usually below eye level, and the backward recline of our car seats forces us to turn our eyes down, even when looking ahead along the road.

This constant looking down, in time, affects the eyelids. The lower lid being constantly turned down becomes creased and baggy, and may eventually fall away from the eye — a condition known as ectropion. The lubrication of the eye, instead of being swept across the eye and down into the nose, collects behind the sagging eyelid, until it wells over in tears. To date surgery has been the standard treatment.

Further variations of posture affect eyesight. For example, if the right leg is shorter than the left, then the pelvis is tilted to the right, and the head is tilted to the left to counterbalance the pelvic tilt. The eyes have to be rotated to the right in order to see straight in front.

Right handed people usually tilt the head to the left and rotate it to the left. These alterations are even more pronounced if the person's work requires precision, and the necessity to get the eyes behind the work, for instance in carpentry. To line up their eyes behind their right arm, when it is in action, right handed people tend to turn their eyes to the left and rotate them anti-clockwise, as well as tilting and rotating the head to the left.

On the other hand, deficiencies of eyesight may affect posture. The totally blind hold their heads back, so that the head is less likely to be hurt if they walk into something. With the result that they usually have good posture. If only partially blind, and the best vision is out of one side of the eye — then the head will be turned to bring this side of the eye to the front.

Eye specialists are well aware of the wry neck (torticollis) caused by squint (strabismus). In fact it bears a name, ocular torticollis, and ophthalmologists are often able to tell which of the twelve external muscles of the two eyes is not functioning simply by observing the position in which the head is held. Head posture is an important item in the signs and symptoms of squint, and photographs of children with abnormal head posture are shown in the textbooks which deal with squints. If head posture is distorted, then, for balance, the rest of the child's posture must be distorted. Fortunately, if the necessary surgery is done early enough the head usually acquires a normal position.

A less severe sideways bend of the head may result from astigmatism. In this condition, because of muscular imbalance, the child sees double when the head is straight. By bending the head in a particular way these two images become one.

> **There is a very close relationship between posture and eyesight. Either may affect the other.**

The other eye effects caused by modern life are not postural, yet they deserve mention if we wish to preserve or improve our eyesight. Our ancestors spent much time looking into the far distance to detect the slightest movement or clue, which would warn them of danger, predators, or a possible meal.

Their heads were held mainly in a constant position, with their eyes doing the directional movement, so all their eye muscles became well developed. Furthermore, their food and their very lives depended upon accurate, detailed sight from a distance — perhaps the movement of a few leaves a mile away. To see these minute details either in the distance or close up, a special spot on the retina, called the Fovea Centralis, must be used. The modern equivalent of early man was the wartime pilot or air gunner, whose life depended upon ceaseless movement of the eyes and minute observation for early detection of the enemy.

Today, our eyes look for long periods in one direction, and rarely do they move through their full range, except to look down. We seldom look at minute detail whether looking into the distance or at close work. We see the tree not the leaf. We are taught to see and read the word or even the whole line, rather than a small part of one letter. This means that not only the eye muscles weaken; but we lose the ability to see in accurate detail. Working in poor light takes vision even further away from this central, accurate seeing point, for we use more of the periphery of the retina to see. If we do not use the Fovea Centralis to see in accurate detail, we gradually lose the *ability* to use it.

It is claimed that reading fine print requires just as good a fovea as distant detail spotting. From my desk I can discern a bicycle (two metres) moving on a road two and a half kilometres away. This is equivalent to a single letter in fine print of only 0.27 millimetres in width seen from a distance of 330 millimetres. I could not read such fine print, neither would I get practice in using the fovea for I read whole words not individual letters. Furthermore, my ancestors would have been able to see much finer detail at a much greater distance than 2.5 kilometres.

There is another great weakness in this comparison. In reading small print the eyes have to converge (turn inwards towards each other) quite considerably for vision at reading distance, to bring both eyes on to the subject. In viewing objects at a great distance there is no convergence.

We are so dependent upon other people — shoe manufacturers, furniture, car and industrial designers — that it is almost impossible to use our eyes as they were meant to be used.

But innovation is possible. We can remove the heels from our shoes; raise the television above eye level; tilt and raise the desk and typewriter; make a reading stand, cover up the bottom half of our glasses whilst reading or watching TV, and think of new ways to do routine jobs while looking ahead or upwards.

Such innovation may stop any further deterioration of our eyesight, but to improve it we must strengthen the wasted (degenerated) superior rectus and inferior oblique muscles, — or accept the necessity of wearing glasses. The quickest and most satisfactory way of strengthening a muscle is to work it against a gradually increasing resistance (usually a weight), unfortunately this cannot be done with eye exercises and only slow progress can be made with free exercises. But development will come with daily practice, which must be to the very limit of the range of contraction.

So, if you hate glasses enough and you are prepared to do the necessary work, improvement in vision can be achieved.

The following eye exercises will strengthen the muscles that have been weakened by our lifestyle and poor posture. Try them to music.

So that they work together, the muscle which pulls the upper eyelid up, uses the same tendon and nerve as the muscle which turns the eye up. Thus we can work them together against resistance.

1. Hold your eyelids shut by using your index fingers. Try hard to pull your eyelids open. Pull for a count of two seconds. Repeat ten times.

2. Close your eyes. Place your index and middle fingers lightly on the top of the corresponding eyeball. They will be on a plate of thin cartilage inside the upper eyelid, and the eyeball behind it. Look up as high as possible, lifting your eyebrows, also. Your fingers will feel the cartilage move up. Resist this movement with your fingers, for one second. Look down, then repeat the exercise for three minutes.

3. Focus backwards and forwards from a small, near object to a small, distant object — for one minute.

4. Look at distant, minute detail and force yourself to see more detail — for one minute.

5. Swivel your eyes from side to side, as fast as possible, for one minute.

Repeat Exercises 1-5. Do this routine at least twice a day, to music.

CHAPTER SEVENTEEN

POSTURE IN GENERAL HEALTH

"When you are well hold yourself so."

17th Century Scottish Proverb

We have dealt so far mainly with the mechanical forces which are distorted by bent posture, those that act upon bones, joints, backs, muscles and in pregnancy. There is however, an effect on all the soft tissues of the body. When a tall building is distorted by the subsidence of its foundations, the wiring and plumbing may snap, carpets, doors and windows will no longer fit and the lifts will not work. The distortion of a living body with poor posture, however, is a slow, continuing process and the body endeavours to adjust to it. Nerves, blood vessels and muscles do not snap; they become permanently stretched or shortened. The organs of the body, such as the heart and liver, become either cramped for space, or sag, and stretch with too much space. Just as muscles lose some of their efficiency, when they are either permanently stretched or shortened, so do all the working components of the body.

To find out how much modern life, (and the poor posture it produces), is in fact spoiling our hearts and circulation and reducing the efficiency of other organs of the body, we must compare industrialised people with those who are least touched by modern civilisation. In terms of evolution, the "civilised" are the abnormal. We must however, hurry with our surveys, for the 'uncivilised' will soon be 'civilised'. Even the remotest villages are being reached by radio and television. Since uncivilised people are the norm in nature's design, we should survey them in a study of every facet of human life.

> **Physiology is the study of the normal functioning of the normal human body, and is the base line on which normal health is decided. Physiologists however, have been studying abnormal ('civilised') people, and taking the base line from them.**

The 'normal' pulse rate is said to be 72, but any regular jogger, making some effort to counteract civilisation, knows that it should be much lower than that. The 'normal' level of sugar in the blood is considered to be between 80 to 100 milligrams per 100 millilitres of blood. But is this correct? My investigations suggest that it should be much lower than that. Truly normal values as a measure of health, need to be taken from those uncontaminated by 'civilisation'.

In the history of medicine it is found that the great advances in combating disease were made initially by engineers and not doctors. Provision of a clean water supply, efficient sewerage, and arrangements for garbage disposal eradicated the great epidemics and were the first major advances in health. This is Preventive Medicine and it is now run by Health Departments. More recently, these departments have been concerned with immunising the population against specific diseases. The future of Preventive Medicine lies in the promotion of fitness and efficiency in the individual. Posture is a vital factor. The fight against such degenerative problems as osteoarthritis, cardiovascular disease and back malfunction is the major task which lies ahead of the public health service.

CHEST

The effect of poor posture on the contents of the chest is considerable, because the bony walls confine the available space. If the thoracic spine is bent and the shoulders pulled forwards, the chest cannot expand correctly. If it does not expand efficiently, it will not recoil properly and some of the elastic tissue of the lungs, which produces the recoil, will start to disappear simply because it is not being used. The volume of the chest will be reduced, so that the actual size of the lungs is diminished. These factors lower the overall efficiency of the lungs, so that if disease attacks, defensive mechanisms are less effective.

In such a round shouldered chest, the heart will be cramped for space, be reduced in size, and frequently distorted to one side. The expansion of the chest in breathing not only sucks air into the lungs, but sucks blood along the veins to the heart. Because the return flow of blood to the heart is less efficient, heart function is affected. When blood vessels become shortened or lengthened and stretched, the composition of the blood vessel wall will change. It will have less or more than the right amount of elastic material in it, and its ability to function correctly will be diminished. If the position of the heart is distorted to one side the dynamics of blood flow out of the heart into the main artery is disturbed. The shape and positioning of the valves of the heart will be altered, and the main artery from the heart (the aorta) will have one wall shortened and the other lengthened. Even the coronary arteries, will not function as they were designed, when the heart is markedly displaced by poor posture.

If the curve of an artery is increased by bad posture, for example in the region of the lower part of the spine, then the convex wall of the artery will become thinner and more stretched.

The pressure of the blood stretches this wall further and further until a visible bulge (an aneurysm) is produced. This is a highly dangerous condition, for if an aneurysm is stretched far enough, it will burst — usually with fatal results. It is significant that the common sites for aneurysm are near the curve of the lower part of the spine, at the knee and at the elbow. These are sites where increased curvature from bad posture, or, in the case of the elbow and knee, from muscular imbalance, are likely to occur.

The diaphragm is a dome shaped muscle fixed to the bottom of the rib cage. When the lower ribs are crowded together in the round-shouldered, or splayed out in the hollow-backed individual, and have lost their range of movement, the diaphragm does not move up and down through its full range affecting the efficiency of breathing.

Hiatus Hernia is a hernia of part of the stomach, through the diaphragm into the chest. It is a distressing condition, and it seems probable that as well as factors produced by a wide angle of obesity (Chapter 7) an increasingly lordotic posture will flatten and stretch the diaphragm and stretch the food pipe (oesophagus) both of which conditions are conducive to the production of a Hiatus Hernia.

Conversely the food pipe which passes through the chest to the stomach will become bunched up in the round shouldered kyphotic type. When this happens its walls become slack and fail

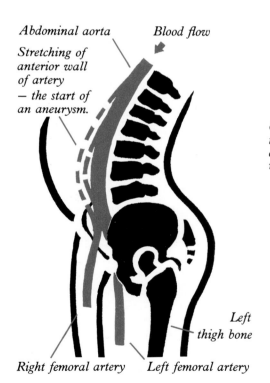

Abdominal aorta

Blood flow

Stretching of anterior wall of artery — the start of an aneurysm.

Left thigh bone

Right femoral artery   Left femoral artery

Figure 17-1. The possible production of an aneurysm (a dangerous bulge in an artery) by increasing the curvature of the artery through bad posture.

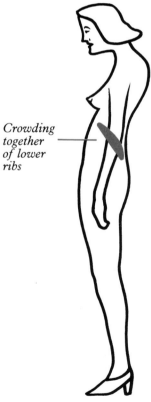

Crowding together of lower ribs

Figure 17-2. Crowded lower ribs of round-shouldered restrict movement and efficiency of diaphragm.

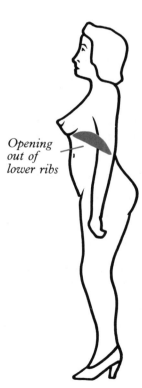

Opening out of lower ribs

Figure 17-3. Splayed out lower ribs of sway-backed restrict movement and efficiency of diaphragm.

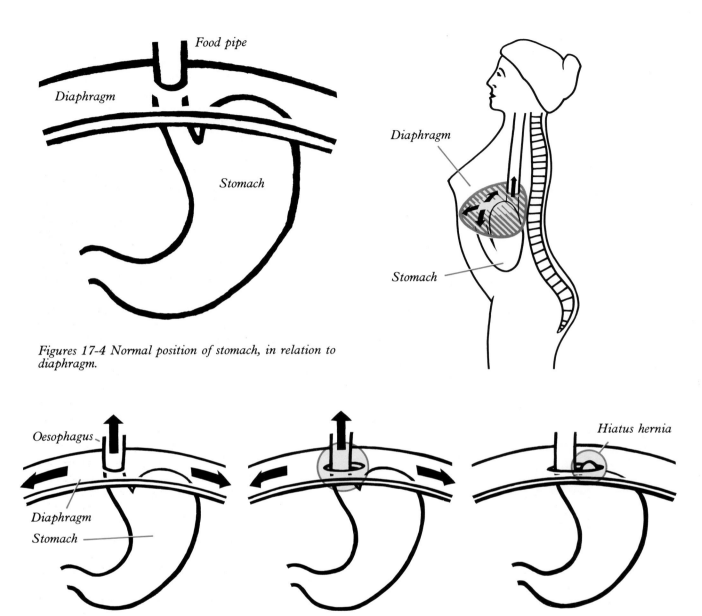

Figures 17-4 Normal position of stomach, in relation to diaphragm.

Figure 17-5 and Figure 17-6. Possible development of hiatus hernia in lordotic person by sideways pull of diaphragm, when it flattens out, and vertical pull of oesophagus from spinal curvature.

to produce the stimulus necessary for the stomach entrance to open sufficiently. The opening becomes progressively narrower. In one humped backed woman the entrance into the stomach had to be forcibly stretched because the food would not pass through. This is usually a recurrent condition in which repeated stretchings are required. This woman improved her posture and no further stretching has been necessary.

There is a condition where the patient complains of continual discomfort and sometimes pain in the lower part of the neck. As they show no abnormality these people are usually labelled as neurotic or as suffering from a psychosomatic condition labelled as 'Globus Hystericus'. In one such woman x-rays showed that the backward curve of the neck was so exaggerated that one neckbone was pushing into the back of the food pipe. After im-

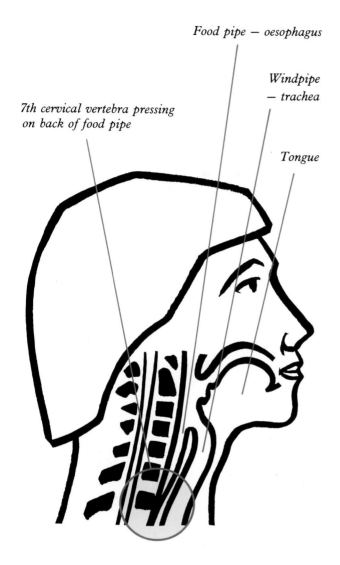

Food pipe — oesophagus

Windpipe — trachea

7th cervical vertebra pressing on back of food pipe

Tongue

Figure 17-7. One of the vertebra of a round-shouldered (kyphotic) person pressing on the back of the food pipe (oesophagus), possibly producing the condition of Globus Hystericus.

proving her posture the symptoms disappeared.

Because the front wall of the abdomen is made of muscle, the contents of the abdomen do not get seriously cramped for space. Abdominal muscles will stretch with poor posture, providing more space than was intended. Most of the organs of the abdomen are suspended by a thin sheet of tissue called the peritoneum. If the posture is poor and the abdominal muscles are stretched and weak, the peritoneum will stretch and the organs sag. I recall one woman who had an operation to investigate a large lump, low down in her abdomen. This was found to be part of her liver, which had slid down a distance of about five inches. No other abnormality was found. This woman was tall, with an extreme stoop. Unfortunately, improving her posture at this late stage did not move her liver, for the tissues had become permanently stretched.

The position of the top of the stomach is fixed, because it is continuous with the foodpipe. But, the rest of the stomach is free to sag and stretch, when posture is imperfect and the muscle wall of the abdomen is weak. The stomach is a dynamic piece of human machinery that is squeezing and churning constantly to digest food, and will not work effectively if it is stretched and distorted in shape. Similar principles apply to the intestines, and particularly those parts of the intestine which are supposed to be fixed in position. With lax, pendulous bowels, constipation, as well as inefficient absorption and digestion of food is likely to occur.

Strong abdominal "stomach" muscles can in fact be used voluntarily to stimulate the bowels to open as with yoga type exercises. There is a condition, diverticulosis, found commonly from middle age onwards, in which small pockets appear in the sides of the bowel. They occur where there are weaknesses in the muscle wall of the bowel, and are said to be the result of constipation. However, it seems more probable that they are the result of a sagging, pendulous abdomen caused by poor posture and weak abdominal muscles. Frequently they cause only minor disturbance, but occasionally an abscess develops in a pocket with dangerous results.

The gall bladder concentrates the bile from the liver before it passes down a tube (the cystic duct) to the intestine, enabling fat to be digested. Millions of years of evolution have determined accurately the size, direction and curvature of the tube to the intestine. If all these factors are changed by incorrect posture, efficiency cannot be expected. For generations, medical students have been taught that gall bladder disease occurs in the fair, fat and forty. The 'fairness' remains unexplained. The 'forty' probably related to the time it takes for a gall bladder stone to grow big enough to cause symptoms. But, the 'fat' unwittingly refers to those with a wide angle of obesity, who tend to become fat. They also tend to be sway-backed, and this is a possible reason why they develop gall bladder disease. The curvature of the sway backed increases the distance between the gall bladder and the point where the cystic duct enters the intestine. A stretched cystic duct inhibits the emptying action allowing bile salts to accumulate.

There used to be a term 'floating kidney', in which the kidneys had drooped down and away from the back wall of the abdomen, and surgeons used to sew them back. This condition, for which surgery is hardly ever used now, is almost certainly a postural abnormality. The two tubes which carry the urine from the kidney to the bladder (the ureters) are prime targets for being misplaced and misshapen by posture.

The urge to pass urine comes when the bladder wall is sufficiently stretched to stimulate the appropriate nerve centres in the brain. If the abdominal muscles are weak and stretched from bad posture, then the bladder wall will gradually stretch and the stimulus for the bladder to empty will be delayed.

The floating kidney, the distorted tubes from the kidney to the bladder, plus delayed emptying of the bladder all increase the likelihood of bladder and kidney infection. If the bladder is tilted forwards by weak abdominal muscles, by overfilling, and by poor posture the prostate, which is a gland encircling the exit from the bladder in the male, becomes distorted and probably enlarged. When this happens passing urine becomes difficult and sometimes impossible.

There is a distressing condition called stress incontinence, when urine is passed involuntarily by women during coughing, sneezing, laughing, running or jumping. It occurs after childbirth, and is the result of the extreme stretching and often tearing of the front wall of the vagina at delivery and the stretching and tearing of the urethra — the tube which empties the bladder. It is usually treated by surgery, which may fail to correct the condition. It is often many years after delivery before the onset of the problem and posture almost certainly plays a part. Two patients whom I treated for osteoarthritis of the spine by correcting their posture, reported that their stress incontinence disappeared.

Joggers have found that when their abdominal muscles become tight from running, an urge develops to pass urine or empty the bowels (sometimes to the embarrassment of the jogger) and most

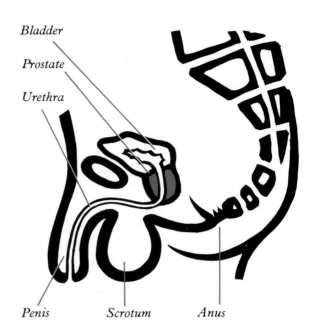

Figure 17-8. Cross section through bladder, prostate gland, urethra and penis.

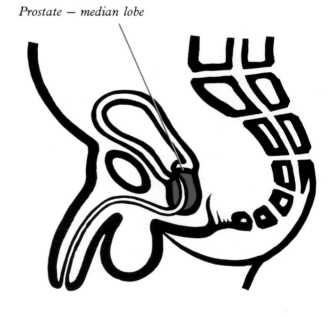

Figure 17-9. Cross section as in 17-8, but in lordotic person, showing how the median (centre) lobe of prostate may become enlarged, so that it is difficult for urine to pass, when the bladder contracts.

would agree that their bowels have worked more regularly and easily since they commenced jogging i.e. since their abdominal muscles became stronger and tighter.

In the majority of women the womb is tilted forwards. In some however, it is tilted backwards (retroversion). Whether this tilt is important, and whether it warrants operation is still open to conjecture. But, the backward tilt is almost certainly due to weak ligaments, themselves the result of poor posture.

Pressure on nerves from abnormalities of the spine causes severe disturbance to their function, resulting in pain and paralysis. It may well be proved that stretching or shortening them also interferes with their efficient functioning, i.e. in conducting electrical current. The two women with the most marked sway-back that I have seen both developed paralysis of the legs, in later life. In neither case was the paralysis fully explained.

A close friend had a bullet through his head in Vietnam, producing paralysis of the left arm and leg. Yet, with immense courage and perseverance he continued with his favourite pastime — weight-lifting. I was able to advise him on exercises to improve his distorted posture, his footdrop and other problems. Early one morning, eleven years after his injury, he woke me with the news that he had moved his fingers. His joy was ecstatic and justified. A few months later his joy was increased when his ability to straighten his arm and do a press up returned and further improvements are still occuring. This is a unique recovery. Stimulus from the constant exercise must be the main explanation, but correction of distorted posture assisted.

Unfortunately, the paralysis of muscles on one side of his body makes it impossible for him to achieve muscular balance, so that he cannot completely eradicate postural deformities, and suffers from back problems. Similar difficulties arise when treating any one with muscular paralysis, such as those who have suffered polio or a stroke.

There is a condition called osteoporosis, in which the amount of material in the bone substance is reduced, so that the bone becomes weak and breaks easily. It is common in women at and after the change of life. It is thought to be the result of a decrease in production of the female hormone, oestrogen, which occurs at this age. When oestrogen is administered to these patients they improve, or, at least, any further deterioration is

arrested. It is known that giving oestrogen to women of this age will increase their likelihood of developing cancer of the womb (uterus). However, women of this age have a tendency to fall and break a hip, especially if the bones are weak from osteoporosis, and fracture of the hip often proves to be fatal. It is considered that the risk from the broken hip is greater than the risk from cancer of the womb. Furthermore, the pain from osteoporosis may be severe and constant. So there is considerable justification for oestrogen treatment, if osteoporosis is truly a hormone deficiency disease.

Fracture of the hips is much more common in women that in men because the hips jut out further in women, and are more likely to take the full force of a fall. But if the hip joint is further bent from wearing heels on shoes, the hip will jut out even further, and the bones and joint are weakened by this flexion (Chapter 10). It is my contention that the removal of heels should have a greater priority than the administration of female hormones or calcium in treatment designed to prevent fracture of the hip.

I believe that osteoporosis is primarily an exercise deficiency and postural disease, not a hormone deficiency disease. Forced inactivity as in the bed-ridden; or weightlessness as in space flight, both produce osteoporosis.

In space flight, the worst affected bones are those of the spine, the legs, and the heel — in increasing order of severity i.e. the bones that normally carry most weight are the greatest affected when the stimulus of weight is removed.

> **The body's principle of economy works in both cases. Good strong bone is not produced unless it is needed to support weight.**

It is known that those with osteoporosis do not eat much calcium-containing food, and that they cannot absorb calcium efficiently from the intestines. It is probable that this is nature's way of regulating the calcium deposited in the bones, when it sees no reason to maintain the strength of the bones. Physical activity normally increases the desire for food containing calcium. In fact, the body often regulates the types of food we eat by regulating our taste. In the tropics, where sweating causes a big loss of salt, salt tastes delicious. In the arctic, it tastes horrible and is likely to produce vomiting. The obsessional food cravings of the pregnant woman are not neurotic. They are highly likely to have been caused by a vital shortage in the body, probably of a trace element. Exercise is known to increase the absorption of calcium from the intestine. It has been shown that the bigger the muscles around the pelvis the greater is the concentration of calcium in the bone. In the big, overweight woman osteoporosis is rarely seen. The weight she carries is enough to stimulate the maintenance of bone strength. With her large angle of obesity she can extract the most from her food, including calcium. It is the thin woman, with the narrow angle of obesity, who commonly suffers from this condition. She has little weight to stimulate thickening in her bones, and has poor food absorbtion capabilities.

Frequently, in osteoporosis, the bones of the spine, especially in the region of the chest, are likely to collapse producing a condition known as a dowager's hump (Chapter 11). The pain of dowager's hump is likely to restrict physical activity, aggravating the osteoporosis. Other postural conditions such as osteoarthritis elsewhere in the spine, or in the hips, or knees will also restrict activity and produce osteoporosis from disuse.

Occasionally there are reports of strokes following manipulation of the neck, or sudden turning of the head. The vertebral arteries run up either side of the neck through holes in the top six bones of the neck (the cervical vertebrae), and then through a hole in the skull to supply blood to the base of the brain.

It is claimed that the stretching of this artery, during rotation of the head, may cause damage to its lining, resulting in a blood clot. If there is poor posture, with distortion of the cervical curve, then this artery is likely to be stretched or shortened even before rotation commences, and if arthritis has already developed, there may be bony projections which restrict or damage the artery.

Many elderly people with poor circulation use several pillows in bed, to aid their breathing. However, if they also suffer osteoarthritis of the neck, the pillows may stretch or shorten the vertebral blood vessels, and with the depressed circulation of sleep, factors are building up to produce a stroke. A nightmare, in which there are rapid and severe

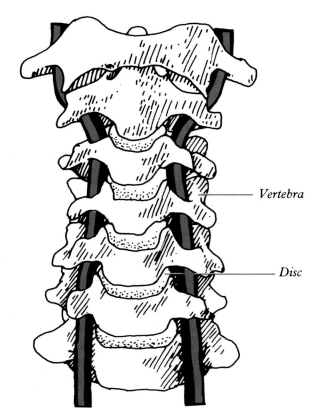

Right vertebral artery and vein

Left vertebral artery and vein

Figure 17-10. Diagram to show how the vertebral artery passes through holes in the top six vertebrae.

changes in blood pressure, might well be the final precipitating factor.

The other artery which supplies the brain (the Internal Carotid Artery) travels up deep inside the neck and lies in front of the bony projections from the upper neck vertebrae. It, too, is liable to be markedly stretched or shortened by excessive curvature of the neck, with possible, long-term, changes in the artery, causing decreased circulation to the brain and a greater likelihood of a stroke occurring.

It is possible that deafness from tinnitus (ringing in the ears) and giddiness are, at least partially, the result of disturbances to the vertebral artery or vein, as a result of distorted curvature of the spine in the region of the neck. By correcting posture I have improved both of these conditions.

In the maintenance of a good circulation it is equally as important for blood to be returned to the heart efficiently, as it is for blood to be pumped from the heart efficiently. The force required to pump the blood through veins back to the heart (i.e. to lift the thick heavy blood up from the feet to the heart) comes from the pressure of the pumping heart; the compression of veins in muscles during contraction (particularly the calves); and the suction produced in the chest, on inhaling.

It has been shown that poor posture interferes with the shape and size of the chest, and on the movements of the ribs and the diaphragm, reducing the efficiency of breathing and also of the body's ability to suck blood back into the chest and heart through the veins. The contraction of calf muscles plays an important part in the circulation of the blood. In fact the calf muscles have been referred to as the second and third hearts of the body. If heels are worn on shoes the range of contraction of the calf muscles is considerably reduced, lowering the efficiency of the circulation and increasing the tendency towards swollen ankles.

In a person whose circulation has deteriorated for other reasons, this further reduction in efficiency may be enough to tip the balance and produce heart failure or a coronary attack.

The role of poor posture in the production of disease and bodily malfunction has been almost entirely neglected. Since the correction of posturally induced disease requires almost costless and harmless treatment, it is irrational to give priority to very expensive, complicated and dangerous procedures before the benefits of postural therapy have been thoroughly examined and exploited.

TREATMENT: See Chapter 21 for:
(A) Essential Exercises 1-3
(B) Recommended Exercises 5-10
IN ADDITION:
(C) If Round Shouldered Exercise 4
(D) If Sway Backed Exercises 22, 23

CHAPTER EIGHTEEN
THE AGEING PROCESS

"Old age, though despised, is coveted by all."

19th century proverb.

"I like to keep young, but my age is against me", wrote an eighty-year-old woman. It certainly showed courage, but where posture was concerned, was not accurate. For posture can be corrected at any age. She may have been told that her arthritis was due to wear-and-tear, and that she was like a worn out tyre with the canvas showing — the less used the better. But comparisons between living and man-made machines are invalid. Of course, if you believe that you are too old for exercise, or that you deserve a rest, then age really is against you!

The critical decision is whether you *want* to live to a ripe old age. Middle aged people frequently claim that they do not want to become old, and say that they would hate to live until they were eighty. They usually hold this view until they reach seventy-nine, when they develop an intense desire to reach eighty and beyond. Life is very dear to us. In fact, we all want to live as long as possible, *providing we are enjoying life.* You certainly cannot enjoy life if you are continually racked with pain. Arthritics are often told to learn to live with their pain. They may be able to exist with it, but to live with it is impossible. A few cannot even exist with it, and resort to suicide to relieve the agony.

If we want to live forever, what stops us?

It is generally accepted that we have an allotted span of time, before the knocking off signal. This belief seems too simple for nature's ways. Furthermore the range of allotted time may vary from 50 to 120 years, too great a range for nature's normal methods. There has been evidence put forward to suggest that we have a built-in time clock, which dictates how long we will live. However, the evidence has been obtained in the laboratory, not in the living being, and must be suspect. The time clock theory is totally fatalistic and is not supported by a growing body of evidence which indicates that *man has a substantial degree of control over his ageing process.*

Skin transplanted from an old animal to a young one lived longer and when the circulation of a growing animal was connected to the circulation of an ageing animal, it was found that the ageing animal rejuvenated. It appears that the old brain was recharged by the young animal's efficient circulation. So get your boilers stoked, your turbines turning and recharge your "ageing" brain and body.

There is a straightforward explanation of the ageing process and death.

If we lived in a perfectly harmless environment, free from all stress or injury and if every system, organ, and cell in our body received the perfect level of stimulation necessary for their regeneration, we could live forever! However, damage inevitably occurs that reduces the efficiency of the body, and over the years the total efficiency of the body is reduced by this cumulative damage. Eventually, there comes a time when the body's total efficiency is not able to maintain an adequate circulation to the brain. At this point we die. i.e. *When the total cumulative destructive forces affecting our body, throughout life, have reduced our efficiency below a certain minimum survival level.*

A bullet through the heart is a very destructive force and reduces our efficiency "totally". Destructive forces can be of three types:
(1) *Degenerative processes* usually reversible.
(2) *Stress Diseases* reversible or irreversible.
(3) *External Attack* usually irreversible.

(1) *DEGENERATIVE PROCESSES*

Degeneration occurs when stimulation is reduced. If a muscle is not used it becomes weak (atrophies). If a joint is not used the blood flow to the joint decreases. If a nursing mother stops feeding her baby, her milk supply dries up. If we keep the stimulus constant the efficiency remains constant (all other factors remaining constant). When the stimulus is gradually increased efficiency improves up to the maximum limit for the particular individual, muscles gain strength, circulation improves, or milk supply increases.

To make the lungs regenerate exercise is needed which will force the chest to expand and the lung tissues recoil to expel air. We need to pant, to gasp for air. Then the amount of elastic tissue in our lungs will return. To stimulate the return of elastic tissue to blood vessel walls, so that they recoil efficiently, circulation must be stimulated by exercise. Joints, tendons, ligaments and similar tissues will be corrected by posture exercises and by moving joints through their full range.

The organs of the body including the brain and nerve tissues will regenerate under the stimulus of improved circulation, oxygenation and correction of diet.

> **Degenerative processes are almost always reversible.**
>
> **If degenerative processes were the only factors and we totally controlled them we may well live forever.**

However, if we let degeneration continue unchecked by watching T.V. fifty hours per week armed with cigarettes, alcohol and excessive quantities of food and devoid of exercise the circulation may drop to such a low level of efficiency that a heart attack or stroke ensues and the damage done is irreversible. For when the oxygen supply or possibly the electrical supply to the tissues in question (heart or brain) is cut off for long enough, this tissue is killed and replaced by scar tissue forever. It never reverts to the stuff hearts or brains are made of. The damage and decreased efficiency is permanent. Our society and 'modern' lifestyle is beginning to produce such calamities as early as 20 years of age.

> **In the appendix, I propose a theory (Electric Man) to explain how increased physical activity produces electricity to recharge the brain and operate the body.**

I have also postulated that the division of cells is controlled by electrical forces. Cancer, the major anomaly in the classification of destructive forces (because it is impossible at this stage to accurately categorise it) occurs when cells of the body start dividing abnormally. Cancer, regardless of its cause and the site of its occurrence, would seem to be associated with a localised variation in electrical supply. Evidence suggests that the electrical charge on the surface of a cancer cell is abnormal.

The localised manipulation of electrical forces could well become a major factor in the treatment of cancer.

(2) STRESS DISEASES

Life has always produced problems, anxieties and dangers which induce our brain to signal an increased production of adrenalin. Adrenalin increases the speed and force of our circulation and this increases the electrical supply to the brain (appendix) so that we can make a powerful physical response to our problem. However, in modern life, the frontal lobes of our brains which make our conscious voluntary decisions, frequently decide that we will not make any physical response, but instead, we will remain slumped in our chair. A phone call may set our nerves on edge, and make us pour adrenalin, without being able to make a normal physical response. This leads to an excessive build up of static electricity in our brains — a veritable storm cloud of tension. Unless we "blow our top" we can feel it building up inside us all day. This static usually has to break out somehow with increased physical activity — the cold dinner may be thrown at the husband, the children or the wife may be beaten and battered, the telephone box may be ripped apart, the train load of football fans may try to wreck the train interior, the political rally may become violent. When Mrs Thatcher was British Minister of Education a rally protesting against the withdrawal of school milk became violent. The protesters had marched about half a mile. Some months later a much bigger rally was organised. The organisers routed this march around the streets of London, before heading for Parliament, a distance of several miles. It dispersed peacefully. Presumably the protesters had worked off their steam, their adrenalin, their static electricity, and arrived in a better humour. At the annual Round the Bays run in New Zealand, 70,000 joggers finish a seven mile run at a small park. Although they then search for friends, clothes, drink or food, squabbles are unknown and a more good humoured crowd would be hard to find. Many of the school children of yesteryear had to walk, run, cycle or horseback ride several miles to school, and their evenings and weekends were not spent in motionless hours in front of television. A child was a perpetual motion machine. But nowadays, children are frequently transported to school, where physical activity has been markedly reduced, and usually return to leisure hours of a sedentary nature. No wonder school vandalism and arson is a modern phenomena!

> **As our lifestyle becomes more physically repressed, vandalism and violence increase.**

If we reduce our physical activity to a minimum over a prolonged period, we may even in an extreme case end up with severe anxiety depression and the psychiartrist may prescribe electroconvulsant therapy. I believe that this powerful electrical shock to our brain discharges the built up static causing the muscles to go into violent isometric contraction. Damage from these powerful muscular contractions is prevented by first paralysing the muscles. ECT is very effective treatment! A skipping rope or well-used jogging shoes are, however, cheaper and less dangerous.

While it is very difficult to persuade a severely depressed person to exercise, encouragement to undertake preventive physical activity before this stage is reached, should be the accepted treatment.

Another unnatural and illogical method of counteracting unused adrenalin (to try to prevent depression, high blood pressure and other stress diseases), is to block the adrenalin chemically, by using adrenalin blocking tablets. It seems obvious

> **It is the increased electrical activity of the brain which adrenalin produces, that gives us the CLARITY OF THOUGHT to produce instant physical REACTION, CLEAR THINKING, INTUITION AND GREAT IDEAS.**

that to knock out our most important safety mechanism the adrenal gland, is a highly dangerous procedure, and was certainly not the evolutionary intention.

To reverse or prevent stress diseases we need regular physical exertion. If stress disease continues unchecked, we are liable to experience an irreversible change such as a heart attack, or a stroke.

(3) EXTERNAL ATTACK

When we have an operation, wherever the surgeon's knife cuts, normal tissue is replaced by scar tissue. That's why the scar is with us till we die. If we work in a coal mine we inhale the dust. This irritates the lining of the lungs, and lung tissue is replaced with scar tissue. If we get pneumonia the bacteria destroy some of the inside lining of the lungs, this gets replaced with scar tissue. Anbody who has suffered pneumonia or worked in a coal mine is unlikely to break the world 1500 metre record.

Prolonged exposure to sunshine produces ageing of the skin (a roughening and thickening call hyperkeratosis) and possibly skin cancer.

Car accidents are obviously a major cause of external attack. Broken bones, amputations, torn ligaments and tendons, damaged brains, will reduce considerably our level of efficiency. Smoking and alcohol are now well recognised as dangerous external assaults on the body. Far less understanding and publicity is given to the even greater external attack from cannabis, which concentrates so much on the brain and thought processes. Its toxic properties are cumulative and soon build up high levels in the body.

With technological advance, man is being assailed from all directions — from radiation, food additives, a host of new chemicals in daily use; from noise, air and light pollution. However, technology in the medical field has reduced enormously the external attack from disease. Many illnesses, which used to cause a considerable reduction in our total efficiency, can now be treated so that no permanent damage ensues. Skin may revert to normal, if protected from the sun, and the ravages of alcohol and tobacco can be reversed, if abstention occurs before too much is consumed.

Although most external attack is irreversible, much of it is man-made and preventable.

> **When the total irreversible damage to our efficiency, whether through degenerative processes, stress diseases or external attack, stops an adequate supply of electricity to the brain, then it is time to die.**

This concept of the ageing process may be shown by a balance sheet. I have produced a series of extremely simplistic examples to illustrate the point. No account has been taken of inherited genetic differences which would of course affect the results. If total efficiency is 100% and we *arbitrarily* assume that death will occur at 30% efficiency and percentages are estimated and allotted for various damage, the state of the ageing process can be calculated.

The following examples are *not based on exact research data and are only intended to illustrate the general principles involved.* The factors chosen are obviously selected randomly to show a few of the multitude of possibilities.

Person A (male, aged 60, looks younger, office worker)
Initial efficiency: 100%. Current efficiency: 69%.

DAMAGE		REPAIR:	
Mental stress and lack of physical activity in occupation	8%	Jogged for 10 years from aged 49	7%
Pneumonia (aged 30) with permanent lung damage	10%	Weight lifting exercises for 15 years from age 45	6%
Smoked for 15 years (stopped aged 30)	2%		
Fractured arm (correctly set & healed)	Nil		
Serious attack of hepatitis aged 40	8%		
Industrial and urban pollution	2%		
Dietary imbalances (including too high calorie & protein intake)	6%		
Deteriorating eyesight, increasing injury risk, & reducing exercise activity	8%		
	44%		13%

Total permanent damage   44% – 13% = 31%
Resultant efficiency   100% – 31% = 69%

Person B (male aged 60, looks older, labourer)
Initial efficiency: 100%. Current efficiency: 50%

DAMAGE		REPAIR:	
Car accident (severe whiplash injury, producing osteoarthritis of the neck)	4%	Outdoor labouring occupation throughout life	8%
1939-45 War stress	6%		
Heavy drinking for 15 years	20%		
Smoked 30 cigarettes per day since aged 16	10%		
Kidney abscess	3%		
Arthritic knee restricting normal activity for past 5 years	5%		
Poor diet	4%		
Broken marriage	6%		
	58%		8%

Total permanent damage   58% – 8% = 50%
Resultant efficiency   100% – 50% = 50%

Person C (female aged 53 at death, office worker)
Initial efficiency: 100%. Current effency: NIL.

	DAMAGE		REPAIR
		Operation on heart valve	15%
Rheumatic fever & rheumatic heart disease	30%		
Knife wound to buttock	Nil		
Meningitis	2%		
Below knee amputation left leg for blood clot	15%		
Industrial & urban pollution	2%		
Added strain on heart & circulation from two pregnancies & labours	10%		
Puncture of lung in car accident	36%		
	95%		15%

Total permanent damage   95% - 15% = 80%
Resultant efficiency   100% - 80% = 20%

N.B. The car accident, at the time of the accident, produced 36% disability, resulting in death. 26% disability would also have produced death. However, she could have survived a 25% disability — and after recovery from the road accident, the long term disability from the accident would only be valued at a small percentage, perhaps 2%. A fit person, with only 10% initial disability, could have survived the same accident for the total damage would have been only 46%.

Person D (female aged 93, at death)
Initial efficiency: 100% Current efficiency: NIL.

	DAMAGE		REPAIR
Diphtheria	1%	Cycled 10 miles daily to work for 15 years	10%
Pneumonia aged 25 (i.e. before availablity of sulphonamides & antibiotics)	20%	Housework for family of 8 before electrical aids	10%
Kidney infection	6%	Poverty & heavy physical workload early	
Bronchitis	5%	in life, ingrained the habit of a diet of high	
Varicose ulcer (reducing activity in recent years)	5%	carbohydrate, low protein & high fibre	
Osteoarthritis of neck & fingers, restricting activity.	15%	content	6%
High blood pressure following 7th pregnancy	20%		
Further Bronchitis producing emphysema	20%		
A cold	5%		
	97%		26%

Total permanent damage   97% -26% = 71%
Resultant efficiency   100% - 71% = 29% (producing death)

N.B. A cold would only produce 5% deficiency whilst it was present. After recovery there would be no permanent damage. However, the cold was sufficient to kill her.

Person E (male, aged 29, looks much older, unemployed)
Initial efficiency: 100%. Current Efficiency: 53%.

DAMAGE		REPAIR	
Removal of cartilage following football injury	3%	Close attention to diet & health foods	5%
Aged 21 Fractured thighbone in m/cycle accident with resultant shortening of right leg & producing chronic back pain.	5%		
Cannabis smoker for 7 years (Aged 21 – 28)	15%		
Only 18 mths. intermittent work since giving up University aged 22	6%		
No longer has any communication with either parent	3%		
Previous severe psychosis & anxiety depression	8%		
Broken de facto relationship	2%		
Physical activity minimal	10%		
	52%		5%

Total permanent damage   52% - 5% = 47%
Resultant efficiency   100% - 47% = 53%

Person F (female, aged 29, ex-nurse, housewife & mother)
Initial efficiency: 100%. Current efficiency: 90%.

DAMAGE		REPAIR	
Glandular fever, complicated by hepatitis. Age 22	5%	Work satisfaction & physical activity when a nurse.	10%
Smoked 15 cigarettes per day for 5 years (aged 17 – 22)	3%	'A' Grade squash player	3%
Toxaemia of pregnancy during second pregnancy (aged 28)	15%	Marathon runner	10%
Urban pollution	5%		
High protein intake with tendency to gout	5%		
	33%		23%

Total permanent damage 33% - 23% = 10%
Resultant efficiency   100% - 10% = 90%

These simplistic examples are not meant to portray real life situations in exact detail but to illustrate a basic principle of the ageing process.

One cannot consider the ageing process without taking account of hereditary forces. If we accept that stimulus modifies structure then we must accept that stimulus modifies sperm, ovum, chromosomes and, therefore, hereditary forces. In Darwin's theory of evolution progress depends upon mere chance (genetic mutation). Effort, decision making and free choice are of no importance. But, Darwin also accepted the views of the French zoologist, Lamarck, who believed that effort to improve by one generation resulted in improved efficiency in later generations, and, therefore, influenced hereditary characteristics.

> **Our prospects for a long and happy life depend very much upon the genes we inherit. But they also depend upon the way we live.**

Much of the deterioration in strength, stamina and fitness which is said to be due to old age is actually due to a change in lifestyle. At twenty we may be free to exercise and play sport for several hours each day. At forty, we have so many commitments that we may find two hours a week is as much as we can spare. At sixty when we again have time we are led to believe that we are wearing out rapidly and tend to 'take it easy'. Many who participate in energetic physical activity 'late in life' are amazed at their physical wellbeing and zest for life.

It is claimed that age makes us 18mm shorter by the time we reach 60. Yet, I am 50mm taller then when I joined the Royal Air Force 43 years ago. I am not growing. I am straighter! It is stated that age has already produced degeneration and narrowing of our discs and slumping of our backs by the time we reach 30. If these conditions are indeed present at 30, then it would be wise to get down to the gym and reverse the process — BUT DO NOT BLAME AGE. Another claim is that at 60 a man has half the strength in his biceps as he had at 25. Surely, this strength will depend upon how much use his biceps get, and the only way strength can be equated with age is that we are less likely to give our biceps as much work at 60 as at 25, and that in the intervening 35 years some irreversible damage has occurred to our circulation and thus to the blood supply to our biceps.

Age seems to be unfairly blamed for decreasing sight (see chapter 16), sagging skin (see chapter 11) osteoporosis (see chapter 17), and in this chapter it has been shown how decreasing oxygen consumption and lung capacity is related to external damage, rather than age. Age is considered to be the guilty party in raising a man's speaking voice (e.g. from C to E-flat) by stiffening the vocal cords, and vibrating them more frequently. Amongst factors affecting the voice are lung capacity and elasticity (rebound i.e. "breathing out") flexibility of diaphragm muscle, curvature of the spine in the region of the chest, and especially curvature of the spine in the region of the neck. A man's voice may certainly develop a higher pitch as he gets older, but that does not prove that age has produced it.

Similarly, hearing deteriorates with age — but this does not prove that it is because of age. Maybe the hearing loss is due to decreasing circulation to the ear because of incorrect neck posture, or the hydrodynamics of the inner ear may be distorted by posture.

Age gives a longer time and, therefore, a greater likelihood of external forces producing permanent damage. But, apart from this, it is difficult to find any harm that age does to us, whatsoever.

There are three groups of extremely long living people — one in the Himalayas, one in the Caucasus, and the other in the Andes. These groups have many features in common.
(1)   They live in isolated areas so they avoid many infections, industrial pollution and industrialised stress and tension.
(2)   They get an enormous amount of physical exercise on their mountains, for there is no tougher work than going up.
(3)   In the rarified atmosphere, there is stimulus for the body to improve its circulatory efficiency to cope with the reduced oxygen supply. Athletes sometimes take to the mountains for periods of training.
(4)   Their diet is very low in calories. In fact, one group is classified as being below the World Health Organisation's level for starvation. Many have, apparently, been starving themselves to death for over 120 years. They eat proportionately more fat and carbohydrate than does our society, and proportionately less protein. Carbohydrates and fats are the foods primarily used by the body to supply

energy for activity.

(5)   If sexual prowess is any gauge of longevity, they certainly seem to qualify for their men claim to be still reproducing when they are on the way to their second old age pension!

(6)   Their posture and joints are good. In climbing mountains you lean forwards; in descending you lean backwards, so their joints are flexible and muscle tone balanced.

Our previous examples of the ageing process balance sheet show us that if we can reduce the deterioration side to the lowest possible level and increase the repair side to the highest possible level, we will slow down the process or even temporarily reverse certain aspects of it.

CHAPTER NINETEEN
POSTURE IN OLD AGE

"Better wear out shoes than sheets"

17th Century Scottish Proverb.

The search for the antidote to ageing is as old as history.

It has been found that undernourished mice live longer. Most of us get a bit edgy when hunger pangs start to trouble us. It is hunger which wakes most animals, and gets the early bird out of bed. These caged mice probably live longer because they are more active, not because they are undernourished.

It has also been found that, if you restrict the diet of certain fish and later lower the water temperature, the fish live longer. Well, they certainly cannot turn up the central heating. Nor put on a fur coat. As far as I know they cannot even shiver. But, they can move around more, get some exercise, to keep up their body temperature. This would be their natural response to the colder water. We all respond this way. Again it is the activity which produces the longevity.

When the hundred year olds of America were interviewed one common factor was that they had all spent a very physically active life.

Living things respond to stimulus. There must be something that makes us respond. This "life factor" or "response to stimulus" factor is what distinguishes the living from the non-living.

There is a rare children's disease called progeria in which these children "age" in childhood, and look like little old people and have usually suffered a fatal heart attack by the time they are ten. These children may have a reduced "response to stimulus factor", and it is probable that this factor is the hereditary or genetic factor in ageing.

What we can do to prevent "ageing" is to increase the stimulus to ourselves by increasing both our physical and mental activity.

We are told that we shrink, bend and lose height with age. *This is a myth.* We lose height because of bad posture, through bending at the ankles, knees, hips and along the spine.

If posture is bad when young, the forces of gravity will gradually make the bending greater over the years — just as the leaning Tower of Pisa is increasing its lean. If posture is distorted by muscle imbalance from work or leisure activities, the distortion will also increase with time.

DO NOT accept this loss of height as permanent and irreversible.

There is far too much of a defeatist attitude towards old age. One eighty-year-old advised her niece, "Never admit to your doctor to being over sixty, they won't treat you, you know!"

To regain good posture you must exercise in the opposite direction to the bending.

Initially, you can do free exercises, i.e. against no resistance. However, to pull those tight joints back into place after so many years, you will have to toughen your muscles more than a little. Exercise against resistance, (raising a few pounds of weight), is eventually going to be necessary — especially if full results are desired. Do not let this frighten you. If the exercise increase is sufficiently gradual, even eighty-year-olds can participate. Lifting two pounds is less daunting and more sensible than trying to do press ups, sit ups or touching toes.

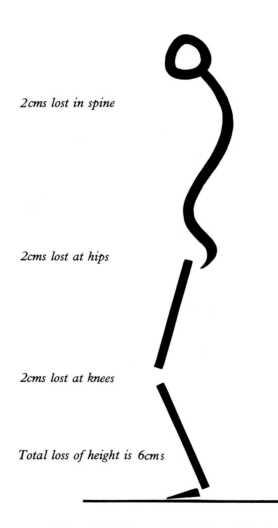

2cms lost in spine

2cms lost at hips

2cms lost at knees

Total loss of height is 6cms

Figure 19-1. Loss of height through bad posture.

For a happy old age, healthy joints are essential. The need to be free of the crippling pain and the immobility of osteoarthritis is obvious. It is wise to make and keep all joints fully flexible. If we do we are less likely to trip and fall and suffer a broken hip — the dread of the elderly. What joy to be able to leap out of bed in the morning free of pain and stiffness. How lovely to be able to bend down and reach up with ease! What fun to be able to keep a flowing golf swing or a smooth bowls delivery. What satisfaction to knit, sew and weed the garden without discomfort! Contrast this with the pensioner who has to work his elbows for ten minutes, then his shoulder for a further ten minutes, before he can use his arms to ease his stiff

back up, to get out of bed. Or to the one who has to hang his clothes up from the ceiling on ropes and pulleys, so that he can slide unaided into them in the morning, having no one to help him.

Provided that your joints have not been permanently damaged by such joint disease as rheumatoid arthritis, you can return them to full flexibility. They may be as stiff as a board, but that's because you have not put them through their full range of movement for many, many years. Persevere and they will be restored. Correct your posture and your weight-bearing joints (those of the knees, hips and spine) will come right. Work the other joints against the bend they have developed and they will also become flexible and pain free.

If there are periods when there is no physical stimulus to the body, then it starts to lose its fitness and flexibility. For the elderly, it is very important to be physically active every day, not only to preserve the body's present state, but to make up for the years of inactivity and stiffening joints.

The elderly are advised, and wisely so, to have a medical check up before they undertake exercise. Since their bodies deteriorate if they spend too much time sitting in a chair or lying on a couch, *it would seem even wiser for them to have a medical check up if they are not going to exercise. Only the young are fit enough to be inactive — and not for long!*

Life is movement — so keep moving!

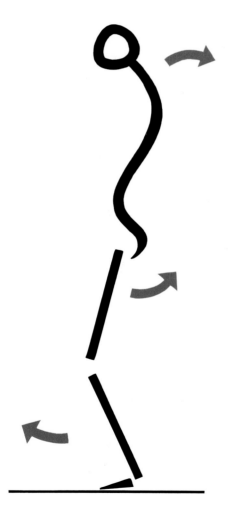

Figure 19-2. Exercise in the opposite direction to bending at a joint, in order to regain good posture.

Treatment: See Chapter 21 for

(A) Essential Exercises 1, 2, 3.
(B) For round shouldered, Exercise 4.
(C) For sway backed, Exercises 22, 23.
(D) Optional Exercises 5-10.
for lordotic or kyphotic.

CHAPTER TWENTY

POSTURE IN SPORT AND JOGGING

"Jog on, jog on, the footpath way,
And merrily vault the stile-a:
A merry heart goes all day,
Your sad tires in a mile-a"

William Shakespeare (1564-1616)

> **The accepted essentials for success in sport are speed, stamina, strength, flexibility and mental approach. To these should be added correct posture and the prevention of postural related injuries.**

During the past eighty years stamina training has been developed and refined in such areas as competitive running, jogging, cycling, swimming, cross country, skiing, and circuit training with or without weights.

The production of strength for sport (a high power/weight ratio) is now moderately well understood and documented. Although, when I first used weight training in 1940 for track athletics it was considered sheer heresy. It is now accepted that, if other factors are equal, the high jumper who can squat with 180 kilograms will jump higher than the one who can only squat with 120 kilograms. What is not so obvious is the amount of control obtained by having a great reserve of power. The strong arm produces the delicate drop shot, the beautiful chip, the graceful leg glide. The strong back brings smoothness and precision to the gymnast and the diver. The weak are unstable in all they do. Even the strongman starts to tremor as his power/weight ratio is extended to the limit.

Skill training and the development of flexibility have been extensively researched and documented.

The mental approach to sports has been (and still is being) extensively researched and documented.

> **But the effect of good posture on movement has been largely overlooked. Posture is the key to further improvement in performance. If posture is perfected, two benefits are derived; the muscular power working on the skeleton will produce maximum efficiency, and the possibility of injury will be reduced.**

It has never ceased to amaze me that so many brilliant sportsmen are forced into early retirement by joint pains resulting from the neglect of their posture.

Muscle pull on bone can be considered in purely mechanical terms. If the chassis of a car is twisted, its performance is reduced and it becomes more dangerous to drive. In the same way, sportsmen and joggers suffer decreased performance plus the liability of muscle tear, tendon tear, joint and back pain, if their posture is distorted.

In discussing backs we saw that the maximum weight that could be lifted in a two pole system was when the poles were almost in a straight line.

Similarly, the most efficient driving force from movement at a joint, comes when the bones of the joint are approaching a straight line.

If there is an imbalance of the muscles acting on a joint (e.g. the flexors are much stronger than the extensors) efficiency will be reduced. The stronger flexors will hold the joint in a flexed position, so that the joint capsules and ligaments on the side of the flexors will become shortened, thickened and tight. Then, when the extensors contract, unnecessary force will be required to stretch the shortened capsule and ligaments. They may be so tight that full straightening of the joint is impossible, reducing efficiency even further.

In competition, the weaker muscle acts against both its stronger antagonist and the thickened capsule and ligaments. Tears of muscle, tendon, joint capsule and ligament are likely to ensue.

In the lordotic type, injury and loss of efficiency are most likely to occur around the knee, hip and lower back regions. In the kyphotic, the most likely regions to be effected are the neck, thoracic spine and shoulder.

All will lose efficiency and increase the risk of injury by using shoes with heels.

Some may claim that in trying to change posture, style will also be changed, and that if you try to change the natural style, you ruin the sportsman. If the sportsman had to hold his posture consciously in a different way, there would be truth in this allegation. However, when posture is perfected style changes are subconscious. By changing the relative strengths and, therefore, the tone of different muscles, both posture and style change with no conscious mental effort.

In any sport that involves running, one of the results of correcting posture is an automatic increase in the length of the stride. The postural exercises (Chapter 21) will increase the strength of

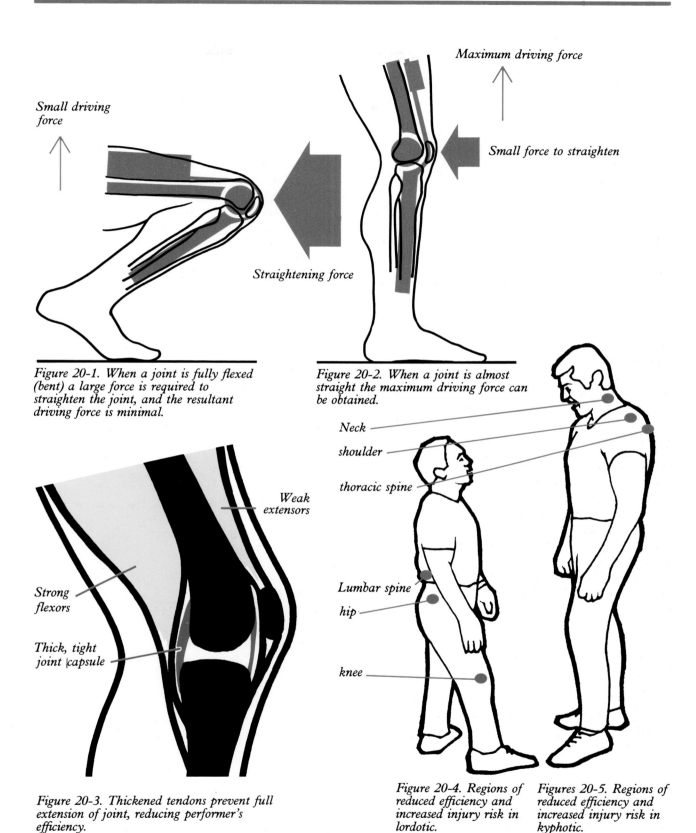

Maximum driving force

Small driving force

Small force to straighten

Straightening force

Figure 20-1. When a joint is fully flexed (bent) a large force is required to straighten the joint, and the resultant driving force is minimal.

Figure 20-2. When a joint is almost straight the maximum driving force can be obtained.

Weak extensors

Strong flexors

Thick, tight joint capsule

Figure 20-3. Thickened tendons prevent full extension of joint, reducing performer's efficiency.

Neck

shoulder

thoracic spine

Lumbar spine

hip

knee

Figure 20-4. Regions of reduced efficiency and increased injury risk in lordotic.

Figures 20-5. Regions of reduced efficiency and increased injury risk in kyphotic.

the muscles which extend the knee and hip joints. This means that the knee of the driving leg will fully straighten and drive over a longer range and the hip will extend further enabling the driving leg to work through a wider range.

Joint mobility is vital. But to forcibly stretch a joint can be counterproductive.

If you wanted to increase the mobility of the left wrist joint, making it possible to extend the wrist backward further and you use the usual method of pushing the fingers of the left hand backward, using the palm of the right hand, this will force the wrist backward. When it starts to hurt, the stretched gripping muscles and tendons go into violent contraction, to prevent further stretching and protect the joint, the tight joint capsule, and the tight gripping muscles and tendons. This defensive mechanism is known in physiology as the "Stretch Reflex". By forcing the stretched muscles to contract powerfully, they are strengthened and their tone increases. This stretching manoeuvre may even reduce the flexibility of the joint.

> **Although it is possible to overide the Stretch Reflex, the best way to increase flexibility in a joint is to exercise it, particularly with weights, in the direction in which you are trying to increase mobility.**

If you want to increase the ability of the wrist to bend backward, you should exercise in this direction. As you contract the muscles which extend the wrist, a simultaneous message from the brain orders the tight gripping muscles to relax. The tone of the extending muscles increases and extension becomes easier over a wider range.

Generally speaking, to increase flexibility and the range of movement, sportsmen should strengthen those muscles opposing the ones used in the performance of their events.

Increased range of joint movement produces more power, for two reasons. Force can be applied over a wider range for a longer period and muscles

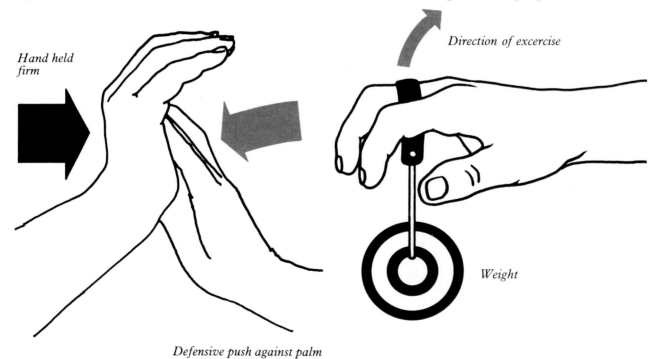

Hand held firm

Direction of excercise

Weight

Defensive push against palm

Figure 20-6. The stretch reflex.

Figure 20-7. Increasing flexibility (extension) of wrist by exercise, as opposed to stretching.

when fully stretched produce a more powerful contraction.

For example, if you increase the flexibility of the ankle by strengthening the muscles which pull the foot upward towards the front of the leg, the Achilles tendon and the calf muscle will be stretched and capable of a more powerful contraction over a wider range.

This drive off the back foot is vital to many sports, it gives the explosive acceleration necessary in all forms of football and hockey, speed around the court in court games and leg drive in field events.

Perfect posture and the flexibility which usually goes with it are essential for all jumping movements. Our tree hopping ancestors are wonderful examples. Before a major leap, a monkey is almost fully bent at the hips, knees and ankles. When it straightens these joints, it gets maximum drive, which is simply maximum force acting over the greatest range, enabling it to make enormous leaps. Before leaping, a cat coils itself up like a spring. Human jumpers endeavour to do the same, but anatomically cannot achieve the same degree of flexion. The jumper is taught to lower the body before take off, increasing the bend of the hip, knee and ankle joints. This is more easily performed if the joints are correctly aligned by good posture, and are flexible.

When the jump takes place, the weight bearing joints are extended to produce the drive and lift. If there is poor posture, full extension will not occur at all joints, power will be lost, and injury more likely.

If all the weight-bearing joints are not straightened in a co-ordinated fashion, the lifting force is reduced and an unwanted rotational force may be produced. The jumper with poor posture must lose efficiency and have to modify his technique to prevent rotation. Coaches would be wise to correct posture early in training, before modified, incorrect, technique becomes ingrained.

Similarly, on landing, the joints must give evenly in a straight line in order to avoid injury. In ballet, where apparently effortless leaps, bounds, rebounds and safe landings on a hard, often sloping surface are a major part of the daily training and performance, it is not surprising that the knee bend (plié) forms the basis of training. Much attention is devoted to posture and drive straight through the centre of gravity. Shoes are light, absolutely flat and as flexible as a glove.

> **Joggers are now training harder than most long distance runners did prior to the 1950s and we must consider them in the same category as the competitive athlete.**

Less downward drive producing less thrust, when the ankle has lost flexibility.

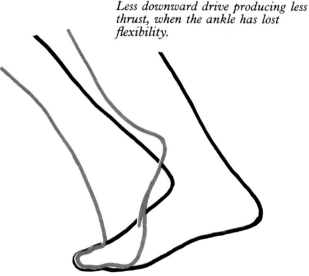

Figure 20-8. With poor ankle flexibility, there is weak downward drive producing thrust, acting over a short period.

Maximum thrust over full range, when ankle flexibility is maximum.

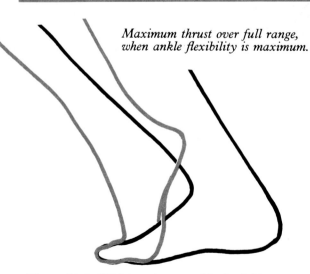

Figure 20-9. With maximum ankle flexibility, the thrust is maximum and acts over a maximum range.

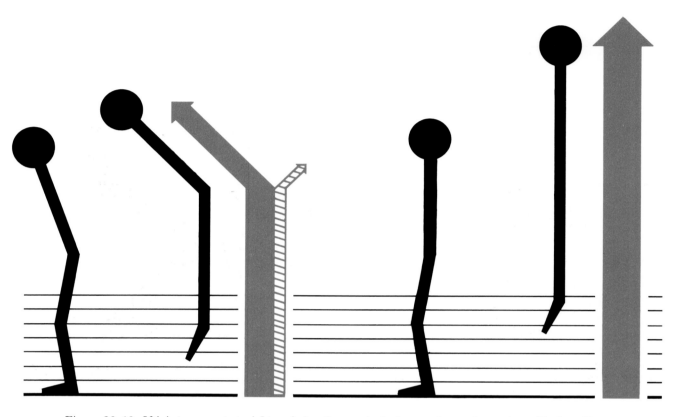

Figure 20-10. If joints are not straightened simultaneously in jumps, forward rotation of body will occur.

The benefits to world health of the jogging phenomenon are of such importance that it would be a tragedy if the "injuries" being encountered were to hinder the continued growth of this activity — especially when these "injuries" are so easily preventable. The problems which distance runners are experiencing are pains and injuries to the knees, ankles, hips, back, shins, achilles tendons and calf muscles. If these problems are allowed to continue untreated, osteoarthritis is likely to develop. These troubles are said to arise from the jarring of the joints caused by continual pounding on hard surfaces. Yet, the human shock absorber system is excellent — far superior to that on the landing gear of a jumbo-jet. It consists of the heel to toe roll on the outside of the foot; the roll from the outside to the inside of the foot; the give at the small joints in the foot; the give at the ankle, knee and hip joints; the give in all the joints in the spine; the compressive qualities of the cartilage in all these joints plus the discs in the spinal column.

If reference is made to the two pole lifting system (Chapter 13), it will be seen that the shock absorbing abilities of a joint are greatest when the bones are almost in a straight line.

The addition of heels to running shoes causes:

- The knees and hip joints to become slightly more flexed, reducing the efficiency of shock absorption.
- The spinal curves to become exaggerated, reducing the shock absorbing efficiency of the spinal joints and vertebral discs.
- The foot flexion to increase at the ankle and the range of joint movement in the "heel to toe" roll to be restricted.
- The stimulus for healthy cartilage and bone regeneration in weight-bearing joints to be reduced and act on the wrong point in the joint and in the wrong direction.
- Direct compression of the cartilage in weight-bearing joints to be converted to a shearing force, which reduces shock absorption and may cause injury to the cartilage.

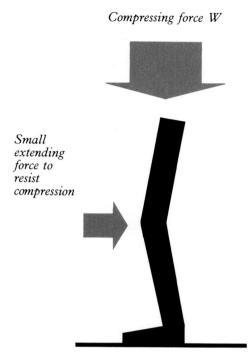

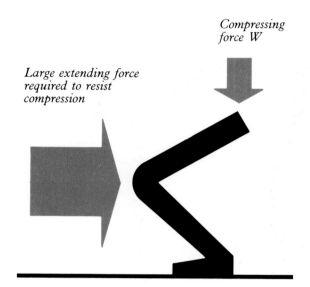

Figure 20-11. Only small contraction force required to resist compression of a joint, when it is nearly straight.

Figure 20-12. Large contracting force required to resist compression of a joint, when it is almost fully bent.

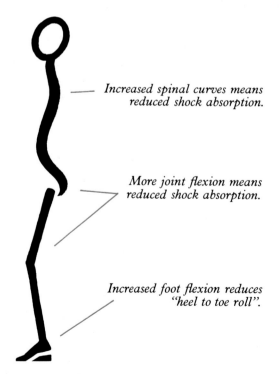

Increased spinal curves means reduced shock absorption.

More joint flexion means reduced shock absorption.

Increased foot flexion reduces "heel to toe roll".

Figure 20-13. The effect of raising the heels on running shoes.

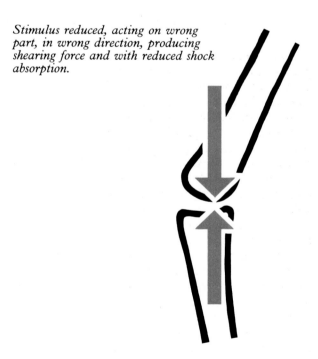

Stimulus reduced, acting on wrong part, in wrong direction, producing shearing force and with reduced shock absorption.

Figure 20-14. Forces acting on knee joint, when heels are added to running shoes.

> **It is ridiculous to spend many hours weekly training or even jogging in unnatural footwear, which diminishes efficiency and increases the risk of injury.**

We need heels on our running shoes like a racehorse needs the back of his racing shoes built up. Our horse would trail the field and almost certainly be pulled up lame. Heels on shoes may be good treatment for achilles tendon or calf muscle injury and plaster of Paris may be good treatment for a broken ankle. But neither is of any use for prevention.

A standard street shoe heel is damaging to posture, joints, ligaments, tendons and muscle in everyday use. The smaller heel on the average road shoe will cause even worse problems when used for over 30,000 strides or more each week. *The achilles tendon and calf muscles get shorter and shorter, and it is patently absurd to spend time with stretching exercises, if our shoes reverse this process.*

Road running shoes should be lightweight, very flexible, cushioned throughout their length, with no heel, but a fine gripping surface curving right back, like an Indian Moccasin. There should be no turn up of the running surface under the big toe. The back of the shoe should not reach as high as the achilles tendon so as to avoid rubbing on the back of the tendon. It is nonsensical to put heels on any shoes used for sports requiring bounding or jumping especially basketball and volleyball. High jumpers are well aware of the mechanical advantages that enable them to jump higher if their soles are thicker (higher) than their heels.

In many sports, there is a one-sided rotation of the trunk. These include golf, baseball, throwing events and fast bowling in cricket. The rotational muscles on one side of the spine achieve far greater

Figure 20-15. The ideal running shoe.

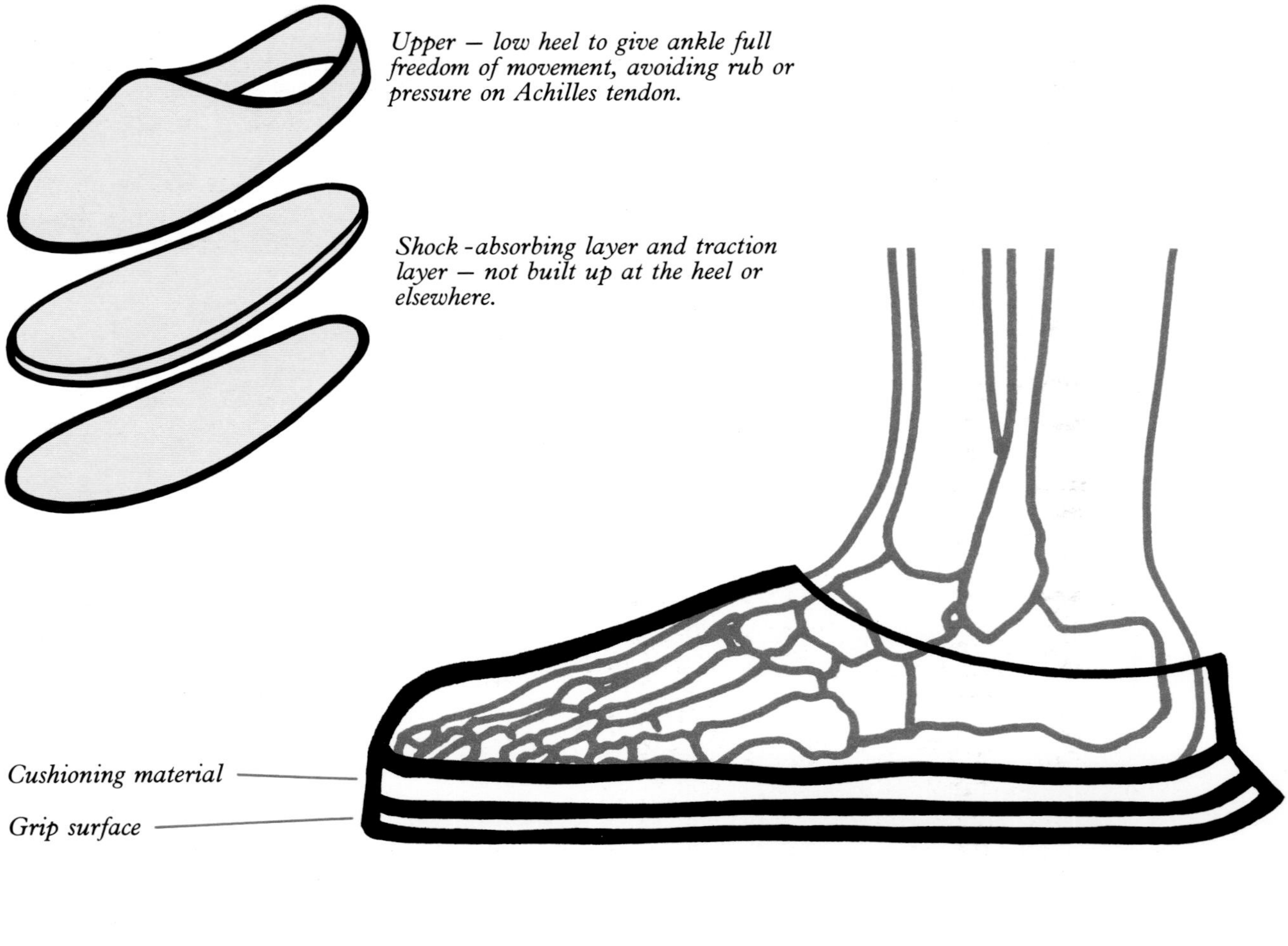

Upper — low heel to give ankle full freedom of movement, avoiding rub or pressure on Achilles tendon.

Shock-absorbing layer and traction layer — not built up at the heel or elsewhere.

Cushioning material ——

Grip surface ——

power than the opposing muscles. Initially, this is of little consequence, but with time, the capsules (the glove-like encasing of a joint) and the ligaments stretch in one direction and shorten in the other. Flexibility is reduced and performance drops. Some of the loss of performance accredited to age is actually due to poor posture and the lack of flexibility that goes with it. This problem has become increasingly obvious with the popularity of Masters (Veteran) sport. It can be corrected completely and if it is not, osteoarthritis is likely to develop.

Other sports require flexion of the major joints and the sportsman is in a bent position e.g. horse riding, squash, tennis, badminton and cycling. If corrective exercises are not done to straighten these joints, performance may drop, posture will deteriorate, injury risk increase, and subsequent osteoarthritis is likely. Because of the excessive development of the gripping muscles on the inside of the thighs (adductors) horse riders often develop osteoarthritis of the hip as well as bandy legs. Jockeys do not suffer in this way because, in riding with a very short stirrup, they rely mainly on balance to maintain their position. In holding back a powerful horse, they develop strength throughout the extensor muscles down the back of the spine, so they have good posture in spite of their small stature.

Squash players seem to be the hardest hit of all court players because they need to bend more when playing. Professional squash is a comparatively new occupation and the early age at which osteoarthritis of the hip, knees and back is developing, is a result of this extreme degree of flexion. The constant bending at the hips, both to strike the ball and to stride forwards or to run to the back of the court, develops the flexor muscles of the hip, so that they are much stronger and shorter than the extensors. At an early age the ability to straighten the hips is lost. The muscles which cross one leg in front of the other are also developed excessively, causing further loss of mobility in the hip joints. If squash players wish to avoid losing flexibility and performance, and to avoid osteoarthritis of the hip at an early age, they must specifically exercise the opposing muscles to the ones developed by their game.

There are sports where head position is vital, such as gymnastics, soccer, skating and diving and this will be ensured by perfect posture. In boxing, head position is important for defence as well as

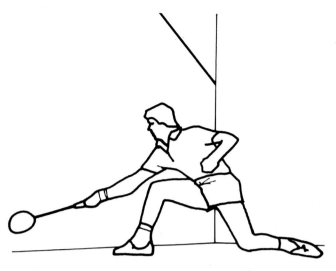

Figure 20-16. The flexed hip position of the squash player.

balance. The boxer with the best professional record must be the ex-light heavyweight, Eddie Cotton from Seattle. Besides his knockout ability, he had an exceptional defence, for he fought 96 professional bouts and approximately 240 fights altogether before he was first knocked down at the age of 41. In my opinion, Cotton possessed the finest posture in boxing, as well as the finest defence.

If you hide your head behind your left shoulder, your head is bent forward to meet your opponent's left jab. But with upright posture, your head is held back, out of reach of punches.

Erect head and neck   Traditional boxer's stance

Figure 20-17. Good posture aids a boxers's defense.

Racquet game players develop a tendency to round shoulders, but as most of these games require stamina rather than great power, the effects are not great; nevertheless, improvement to leverage, length of stroke travel and alignment would be obtained by perfecting posture.

Basketball and volley ball players, because of repeated leaping and stretching usually possess very good posture, in spite of the height of most successful players.

> **It is axiomatic in all sport that the mechanics of body movement should not be distorted by raised heels, or imperfect posture.**

CHAPTER TWENTY-ONE
TREATMENT

"Better to hunt in fields, for health unbought,
than fee the doctor for a nauseous draught.
The wise, for cure, on exercise depend;
God never made his work for man to mend."

John Dryden (1631-1700)

> **REMOVE THE HEELS FROM ALL EVERYDAY FOOTWEAR, EXCEPT FOR THE ODD PAIR KEPT FOR SPECIAL OCCASIONS.**

EXERCISE THERAPY

The treatment of disease by exercise has had very limited exposure, largely because so little serious study has been made of its enormous potential. In fact most doctors have little knowledge of what to prescribe, and there are no medical textbooks on exercise treatment as there are on drugs or surgery. In the index of video and audio tapes, slides and books available from medical libraries to the general practitioner for his further education, exercise rarely warrants an entry, whereas a drug which mimics some of the beneficial effects of exercise is indexed under several separate headings.

Despite the fact that the generally accepted modern treatment of degenerative disease is both inefficient and prohibitive in cost and exercise treatment is cheap and effective, Health Departments and Insurance Companies do not generally promote exercise therapy as legitimate treatment.

> **To be successful the prescription of exercise should be written, specific and detailed with its purpose and performance fully understood by both doctor and the patient.**
> **When correcting posture it is essential to understand which exercises will improve the condition and which exercises will make the condition worse.**

Correct postural exercises are designed to restore nature's balance to the muscles which oppose each other when acting upon a joint.

Treatment must strengthen the relatively weaker muscles and endeavour to avoid any further development of the opposing muscles, so that the imbalance is not perpetuated.

Muscular movements at work, or in sport, may necessitate the use of the stronger muscles and a compromise must be worked out. Where possible, activities or exercises which maintain, strengthen

or further shorten these dominant muscle groups should be temporarily discontinued.

If you are flexed (bent forwards) at the hip, touching your toes is no exercise for you. If you are round shouldered, push ups should be avoided.

The simple solution to all postural and osteo-arthritic problems is to avoid exercising in the direction towards which the joint is bent and to exercise in the opposite direction.

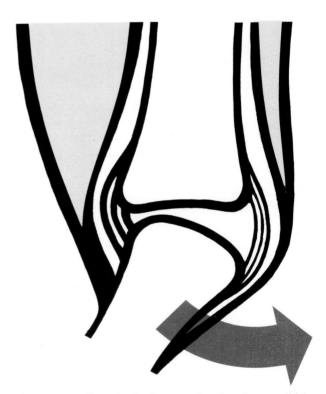

Figure 21-1. Exercise in the opposite direction to which the joint is bent.

When patients of mine have failed to obtain an expected improvement it has frequently been due to their doing additional exercises to those prescribed. They may be exercises advised by the lady next door; seen on T.V. or in another book; advised by physical training experts; just part of their own get fit programme; or because they have always done them. The other main causes of treatment failure are the continued wearing of heels, or the performing of the exercises incorrectly.

Acupuncture, manipulation, drugs, naturopathic remedies, bio-feedback and hypnosis will not interfere with the success of exercise therapy, but fusion of the spine makes subsequent exercise therapy impossible. Artificial knee and hip joints may increase the difficulties in obtaining muscular balance around these joints.

Since our chairborne society generally produces shortened hip flexors and abdominal muscles and weakened, stretched back muscles, postural exercises should generally concentrate on the back until balance is achieved.

Most of my patients were initially in pain, or suffering from some other disability, and many were elderly and frail, so they usually commenced with free exercises (i.e. exercises without any added resistance). Some were quite satisfied with the improvement they made on this regime, and I advised them to continue exercising twice daily for three months and then three times a week to maintain mobility and muscular balance.

Usually, after one or two weeks, the severity of the exercise was increased by the addition of resistance in the form of a weight. Prejudice against the use of weights still exists, but lifting weight is an extremely accurate way of measuring and increasing the stimulus to muscles. It can be easily misunderstood. Lifting five pounds to a height of one foot is weight lifting. So is lifting one ounce to a height of two inches. Not that frightening really?

Weightlifting as an exercise or sport would seem to be particularly free from injury production or circulatory damage.

Only in severe cases of heart or lung disease did I consider that it was unwise to use weights. Usually even the frail and elderly could manage 1 kg.

When a joint has been in a badly aligned position for some time, changes occur in the joint capsule. The capsule on the side to which the joint is bent becomes shortened and thickened, whereas the capsule on the other side of the joint becomes stretched and thinner.

To align the joint correctly, the thickened, shortened part of the capsule has to be stretched, whilst the opposing part of the capsule has to be shortened and thickened — until both sides are in balance. An increase in the tone of the weaker muscle group must be developed to achieve this correction of the joint capsule. A similar process is necessary to correct imbalances in the ligaments around the joint.

Fortunately the muscular imbalance of the

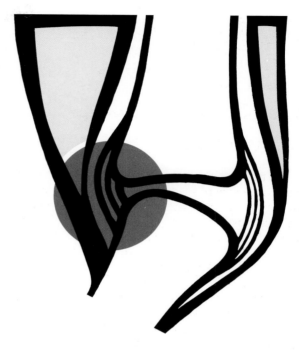

Figure 21-2. The thickened shortened ligaments on the side towards which the joint is bent, have to be stretched, by exercising in the opposite direction.

weak and the frail is easy to correct, so for these people, only a very light weight is required or advised. If the weaker muscle acting on a joint can pull with a force of five kg. and the stronger with a pull of seven kg. then the weaker muscle has only to improve its pull by two kg. for the imbalance to be corrected. Furthermore, the joint capsules and ligaments in these people are weak so that any shortened joint capsules or ligaments can be easily stretched. Frail people respond rapidly to treatment, usually within two weeks, and a maximum of two kgs. weight is often sufficient.

On the other hand, powerful athletes may have quite a considerable difference in strength in opposing muscles. Obviously they need to build up to using heavy weights, and may need several months to show real improvement.

Most patients obtain relief of symptoms and increase their joint mobility with free exercises before they start to use weights, and before complete muscular balance is achieved. The home exercise programme, detailed in the training schedule should be done twice a day. The number of repetitions for each exercise is moderate, as the aim is to build up both the tone and strength of the muscle, not just its stamina. The optimum weight to be reached will depend upon the individual, the particular muscles being strengthened, and the degree of imbalance to be corrected.

AGE IS NO BARRIER TO EXERCISE.

All exercise should be done before meals. After meals, there is a big increase of circulation to the stomach and intestine, and a subsequent reduction of the available blood flow to the heart, muscles and brain.

Many authorities advise a medical check up before embarking on an "exercise" programme. Since most heart attacks and strokes occur at rest and since lack of stimulation produces degeneration, other authorities, more wisely, advise a check up before embarking on a "take it easy" programme. Any athlete will confirm that the reserve of fitness soon drains away if stimulus is not kept up. Only the young and the fit can safely afford to "sit around" and then not for long. Nevertheless, if you are under treatment from your doctor, particularly for heart, circulation or breathing problems, his advice must be sought.

If you are advised against exercise, ask your doctor which will be worse for your health — the stress of light exercise or the degeneration of "no exercise".

There are three basic exercises designed to ensure postural perfection so that the weight of the body and any weight that is carried is borne as much as possible by the skeletal structure rather than by the muscle structure.

The whole weight-bearing system must be corrected at the one time.

The first exercise is designed to straighten and lock the knee, bringing the thigh and lower leg bones into line. The second thrusts the hip forward lining up the thigh bone with the pelvis. The third is designed to correct either excessive or reduced curvature of the spine.

As home exercises these three basics should be done twice a day. If you are training in a gymnasium, these exercises should be incorporated into your programme as your three major exercises. On the days that you do not train in the gymnasium these three basics should be done twice at home.

Exercise only to point of pain.

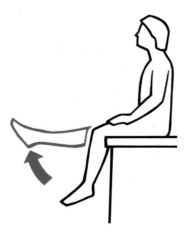

Figure 21-3. Knee straightening.

Exercise 1: KNEE STRAIGHTENING

(1A) AS A HOME EXERCISE WITHOUT RESISTANCE

Method

Sit on a table or bench so that your feet are clear of the floor.

This exercise cannot be done in a chair.

The elderly should take great care in climbing on to the table or bench and may need the assistance of another person. Sit so that the edge of the table or bench is against the back of the knees.

Your hands should rest lightly on the tops of your thighs to ensure that the thighs are not raised off the table during the performance of the exercise. Sit so that your back is upright, but not tense.

Raise your foot as high as possible by straightening the knee. Hold the position with your thigh muscles tightened and then slowly lower your foot. The movement should be done at a reasonably slow pace, approximately one second coming up, one second held, and one second going down.

As the exercise is designed to strengthen the quadriceps muscles (those on the front of the thigh) the foot can be held in the most comfortable and natural position for you.

Repeat the movement until you have done the necessary number of repetitions for one leg, as in the schedule provided.

Complete the exercise with the other leg.

Important Points

With back injuries, this exercise may be painful either in the back or down the leg. In that case, arrange the table or bench so that you can lean back against a wall, when some or all of the pain will be relieved.

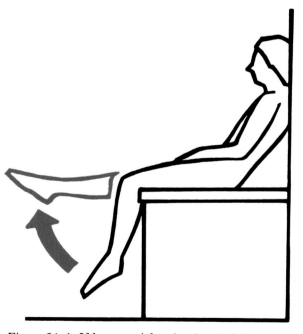

Figure 21-4. If knee straightening is painful lean back against a wall.

If pain still persists, or if you cannot arrange to lean back against a wall, straighten the lower leg only to the point of commencement of pain.

In a few days you should be able to straighten the leg fully without pain.

You must not raise the thigh during the exercise or you will be incorporating a hip flexion exercise, which will counteract Postural Exercise No. 2. This is the reason why your foot must be dangling clear of the ground at the start of the exercise.

Schedule (1A)

> **If you are elderly or in severe pain, follow this programme twice daily.**

First Week:
Eight repetitions with one leg, followed by eight repetitions with the other. Rest for about 30 seconds. Then six repetitions with each leg. Rest for 30 seconds. Then 4 repetitions with each leg.

Second Week:
10 repetitions with one leg followed by
10 repetitions with the other
Rest 30 seconds
8 repetitions with each leg
Rest 30 seconds
6 repetitions with each leg

Third Week:
Most people are, in fact, fit enough to commence with the schedule for the third week, as leg muscles are very powerful.

12 repetitions with one leg followed by
12 repetitions with the other
Rest for 30 seconds
10 repetitions with each leg
Rest for 30 seconds
8 repetitions with each leg

At this point everyone should be able, if they wish, to progress to doing this exercise with resistance.

(1B) AS A HOME EXERCISE
WITH RESISTANCE

Some method of attaching weight (resistance) to the ankle or foot must be used.

I usually prescribe a strapping system designed to hold weight-lifting weights. It consists of a main leather strap approximately 50cm long by 3cm wide, to which is attached at right angles, and at the same level, two side straps. The side straps each approximately 45cm by 2cm secure the weight to the main strap, which can be fastened round the ankle. Any weight up to about 10kg can be attached to the ankle in this way.

An alternative is to use one leg of a pair of old panty hose. 40cm from the toe, tie a knot. Fill a plastic bag with the appropriate weight in sand, soil, lead sinkers or any other commodity, and push this bag down the leg of the panty hose to the knot. Tie another knot to secure the plastic bag in position. 40cm from this second knot cut across the panty hose. Now, using the 40cm lengths either end, tie the weight to the ankle.

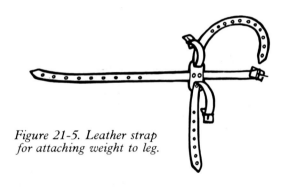

Figure 21-5. Leather strap for attaching weight to leg.

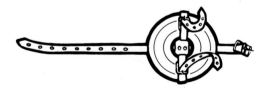

Figure 21-6. Strap and weight attached.

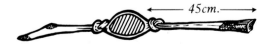

Figure 21-7. Old panty hose used for attaching weight to leg.

Weighted anklets can be bought, but the weight cannot be varied. Iron boots, obtainable from sports shops, are designed to fix varying weight to the feet. However, it is difficult to stand or walk wearing them, so they are dangerous for the elderly or arthritic.

Method

Exercise 1B is performed identically to exercise 1A, but with the addition of weights attached to the ankles. The exercises should be performed slowly and deliberately to prevent the weight banging on the side of the leg.

Schedule (1B)

The repetition pattern should first follow the twice daily 8 then 6 then 4 pattern described earlier. As the exercise becomes easier, progress to the 10 then 8 then 6 repetition pattern and finally to the 12 then 10 then 8 repetition pattern. When this becomes reasonably easy — increase the weight.

(C) AS A GYMNASIUM EXERCISE

Use the leg extension machine to perform the sequences of repetitions described previously. Begin with a moderate weight and progress gradually as described in Schedule (B).

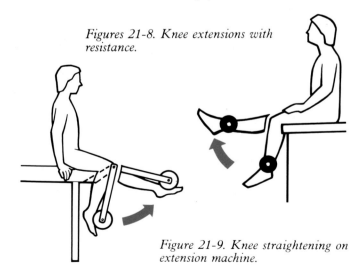

Figures 21-8. Knee extensions with resistance.

Figure 21-9. Knee straightening on extension machine.

EXERCISE 2: HIP STRAIGHTENING

(2A) AS A HOME EXERCISE WITHOUT RESISTANCE

Method

Remove your shoes if they have any heel whatsoever. Lean over a table or bench, supporting the weight of your trunk on your forearms. Turn your toes in towards each other, making your feet "pigeon-toed". Raise one leg up behind you as high as you can, keeping the toes turned in all the time. Then lower the leg to the ground, taking about two seconds to complete the exercise. Breathe in as you raise the leg, breathe out as you lower it, trying to keep both knees straight throughout the exercise.

The aim is to strengthen the Gluteus Maximus muscle (the major muscle making up the buttocks) so the ankle can be held in the most comfortable and natural position for you providing your toes are turned in.

Repeat the movement until you have done the necessary number of movements for one leg. Complete the exercise with the other leg.

Important Points

Rest on your forearms, not your hands. If you rest on your hands the trunk is liable to move up and down as you do the exercise. When you raise your leg behind, you will feel an immediate tendency to twist your hips and spine. If this happens, instead of raising the leg directly behind you, you will actually be rotating the spine and lifting the leg sideways from the hip, using all the wrong muscles. If you remember to keep your toes turned in throughout the exercise, this mistake will be avoided.

Those with severe back pain, who are very stiff, may be unable to bend forward at the hips initially. An alternative method of doing this exercise, for these people, is to stand with the chest against the edge of a half open door. Grip the handles on either side of the door. Turn the toes in towards each other, making the feet pigeon toed. Raise the leg behind keeping the knees straight. The amount of movement obtainable is only small, up to 30°.

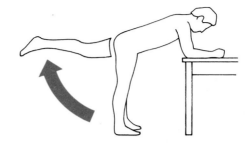

Figure 21-10. Hip straightening, without resistance.

Figure 21-11. When bending of the hip is painful, hip straightening can be performed holding either side of a door.

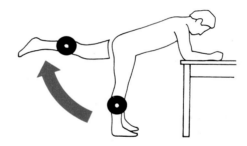

Figure 21-12. Hip extension with resistance.

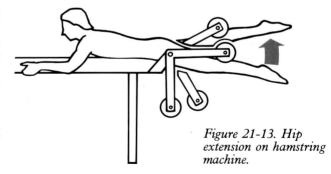

Figure 21-13. Hip extension on hamstring machine.

When there is sufficient mobility of the hip joint, do the exercise leaning over a table, as originally described.

Schedule (2A)

Follow the instructions in Schedule 1A. As this exercise is more strenuous than Exercise 1A those with severe heart or lung disease should commence with one repetition in the morning and one in the afternoon. Caution is the important word in the intial stages. Rapid progress can usually be made after a cautious start.

(2B) AS A HOME EXERCISE WITH RESISTANCE

Attach the leather strapping or panty hose as described in Postural Exercise 1B, to the leg just below the knee.

Method

Exercise 2B is performed identically to Exercise 2A, but with the addition of weight attached to the leg just below the knee.

Schedule (2B)

Most people should start with one kilogram attached to the knee. The more powerful may start with up to five kilograms. It is much more important to do the exercise absolutely perfectly, than to struggle with too much weight so that the exercise is done incorrectly, with the wrong muscles being exercised.

The weight can be increased periodically, but is limited by the difficulty of attaching the weight to the knee. The exercise can be made more strenuous by attaching the weight to the ankle rather than the knee. The exercise should be performed slowly and deliberately to prevent the weight from banging against your leg.

Follow the Schedule outlined in (1B)

(C) AS A GYMNASIUM EXERCISE

Only the best equipped gymnasiums have a machine specially designed for this exercise. However, the machine usually used to exercise the hamstrings can be adapted as a hip straightening exerciser.

Lie face down on the bench as if to do a hamstring exercise, but with the knees instead of the heels under the padded bar. Lift the thighs backwards as high as possible to raise the bar.

Schedule (2C)

Follow the same sequence of repetitions as Schedule (1B).

Begin with a light weight, and increase the weight when a 12, 10, 8 pattern of repetitions can be performed with ease.

Exercise 3 : SPINE STRAIGHTENING

(3A) AS A HOME EXERCISE

Method

Remove shoes, if they have any heel whatsoever. Stand with your back against a wall and your heels about 3cm from the wall. Bend your head forward as far as it will go. Keeping your mouth tightly shut, pull your chin back towards your neck, so that it is almost resting on your chest bone. This position is vital to the exercise and your chin must be held against your chest, with your mouth shut, throughout the exercise. Place a piece of paper under your chin and hold it there by pressure from your chin, and you will maintain the correct position.

Lock your hands together and place them on the back of your head.

Now pull down with your hands and backwards with your head, until your head comes back as far as it will go with your chin still tucked in.

Of course, if the pull of your hands is greater than the pull backwards of your neck, there will be no movement. Reduce the pull of your hands accordingly, so that your neck can be pulled back. Do not let your buttocks come away from the wall, or your mouth open.

Unless your posture is good, do not expect to reach the wall, initially. Stop when your head refuses to go back any further without your chin

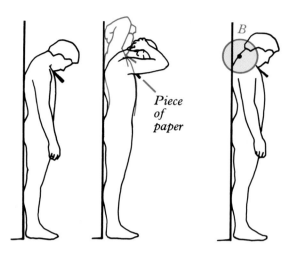

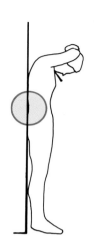

Figure 21-14. Spine straightening.

Figure 21-15. Try to ram the chin towards point B.

Figure 21-16. Try to ram the small of the back against the wall.

Figure 2-17. Spine straightening, holding a book in position in the small of the back.

coming out. All the time you should be trying to ram your chin into point B.

Even if your neck seems rigid when you start this exercise and you cannot get your head to within six inches of the wall, do not despair. Within one month your head should be reaching the wall.

To perfect the mobility of the spine, try also to pull the tummy in and ram the small of the back against the wall. You should be able to reduce the gap between the small of your back and the wall to about 3cm.

Start with the whole trunk bent forward. Uncoil the spine so that the small of the back is against the wall. Try to keep the small of the back pressed against the wall. Practice this exercise by holding a book of about 3cm thickness in the small of the back as well as the piece of paper under your chin.

Important Points

It is very important, but difficult, to breathe during this exercise. The tendency is to hold the breath throughout, which may cause giddiness, a shortage of oxygen, and a rise in blood pressure. If you have this difficulty, stop two or three times during the exercise to breathe. Each repetition should take approximately six seconds, plus additional time for breathing stops.

Towards the end of the exercise, there is often a real urge to tilt the head back on the neck to reach

Figure 21-18. For the frail and those with poor balance, who have to exercise without supervision, the spine straightening exercise may be performed lying flat.

Figure 21-19. Tilting the head back allows the chin to come up. This is WRONG.

the wall, instead of pulling the whole neck back. This should be resisted, as it allows the chin to come up and uses the wrong muscles. The piece of paper under the chin is designed to prevent this cardinal error. Do not neglect it!

In frail, elderly people, especially if they have a humped and bent back, this exercise will be difficult, and possibly dangerous, especially if they live alone and have to do it without supervision. They may topple and fall, one leg may give way, or they may become giddy. If they cannot exercise under supervision, they should commence by doing the exercise lying flat on their backs on their beds or on the floor, and change to the wall when they have improved.

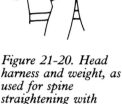

Piece of paper

Figure 21-20. Head harness and weight, as used for spine straightening with resistance.

Figure 21-21. Spine straightening using weights.

Schedule (3A)

The elderly, or those in severe pain, must take care when starting this exercise. Particularly those with severe shortness of breath, caused by heart or lung disease. For such people, the advice on consulting your family doctor must be taken.

> **For the most disabled, follow this programme twice daily.**

First Week:
Do the exercise once. Rest until fully recovered. Repeat the excercise again. Rest until fully recovered. Repeat the exercise a third time.

Second Week:
Do three repetitions of the exercise. Rest until fully recovered. Do two repetitions of the exercise. Rest until fully recovered. Finish with one repetition of the exercise.

Third Week:
(Beginning schedule for the moderately fit, and for those who are fit, but in pain).
Do the exercise four times. Rest for one minute. Repeat the exercise 3 times. Rest for one minute. Repeat the exercise twice.

Fourth Week:
Do the exercise 6 times. Rest for one minute. Repeat the exercise 4 times. Rest for one minute. Finally 3 repetitions. Increase the repetitions in subsequent weeks.

I consider that there is little benefit to be obtained from increasing the number of repetitions beyond a schedule of 12, 10, 8 repetitions twice a day.

(3B) AS A HOME EXERCISE USING WEIGHTS:

Equipment

Head harness obtainable from most sports shops, and appropriate small weights.

Method

Attach the head harness. Hold a piece of paper under the chin throughout the exercise. Bend over at the waist and rest the hands on a table. Bend the head down. Pull the head up, keeping the chin tucked in to hold the piece of paper.

Important Points

Exercise 3A is usually adequate to strengthen the extensors of the spine. Exercise 3B, however, ensures more rapid progress. The range of movement is very small, provided the chin is kept tucked in, being only 3 to 4 inches.

Schedule (3B)

Most people should start with one kilogram. The more powerful may start with up to five kilograms. But remember, that if you overdo it, you will produce a stiff and painful neck. So commence cautiously, carrying out the schedule twice daily.

First Week:
Do the exercise 8 times.
Rest for one minute. Repeat the exercise 6 times.
Rest for one minute. Repeat the exercise 4 times.

Second Week:

Do 10 repetitions. Rest. 8 repetitions. Rest. 6 repetitions.

Third Week:

12 repetitions. Rest. 10 repetitions. Rest. 8 repetitions. When this becomes reasonably easy, increase the weight.

(3C) AS A GYMNASIUM EXERCISE

Although the position of the head on the spine is so vital to posture and to sport and is maintained by the muscles of the neck, equipment for correct neck exercises is seldom available in fitness centres or gymnasiums. Where it is available, it is normally in the form of a head harness attached through a pulley to a stack of weights.

Schedule (3C)

Follow the same Schedule as (3B), making sure that the same techniques are followed in performing the exercise as have been described in (3A) or (3B).

> **These three basic exercises will produce good posture and correct most posture induced pain and disability. After correction, the practice of these exercises three times a week will keep posture flexible and prevent deterioration.**

SEVEN ADDITIONAL RESISTANCE EXERCISES FOR GOOD POSTURE (Exercises 4-10)

For those aiming to have first class posture and to completely prevent posture related disease, the following exercises are recommended, *in addition to the basic programme.*

These exercises can be done once daily, either in a gymnasium or at home. As a minmum they should be done three times a week. Depending upon the time available, you may add the whole

seven exercises to your basic programme, or any number of the seven you choose. Exercise (4) is the most important.

Exercise 4 requires the use of dumbells and if you want to do it at home, a pair of buckets will make improvised dumbells, water or sand being added to increase the weight.

Exercises 5-10 require the use of barbells.

If your arms are not equal in length, you should use dumbells for exercises 5, 6, 7 & 8.

Schedule

The schedule is the same for each of the seven exercises.

First Week

8 repetitions. Rest. 6 repetitions. Rest. 4 repetitions.

Second Week

10 repetitions. Rest. 8 repetitions. Rest. 6 repetitions

Third Week

12 repetitions. Rest. 10 repetitions. Rest. 8 repetitions.

From then on, use the 12, 10, 8 repetition pattern, but increase the weight when the exercise becomes easy.

Exercise 4: BENT OVER LATERAL RAISE

Method

Stand with feet shoulder width apart, with a dumbell in each hand. Bend forward at the hips until the trunk is horizontal (parallel to the ground) and try to keep the back straight. Let the arms hang vertically downwards from the shoulder towards the floor. Raise the arms sideways in a controlled manner as high as you can. Then lower them, also in a controlled manner.

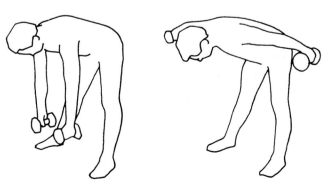

Figures 21-22. Bent over lateral raise.

Important Points

As you raise your arms, do not let them stray back towards your hips. This exercise will flatten the shoulder blades, correcting or preventing "Winged Scapulae", by developing the rhomboid muscles. It is also an important exercise in the treatment of osteoarthritis of the shoulder joint, and in prevention of dislocation of the shoulder, because it develops the posterior part of the deltoid muscles. The exercise should be done at a moderate speed, taking approximately one second to raise and one second to lower.

This is a tough exercise, so start with very light dumbells. If you are a woman and eventually reach 7.5kg in each hand without cheating, you are doing well. (Up to 15kg for a strong man.)

This exercise is identical to the sideways pull with chest expanders.

Important Points

When doing the exercise, concentrate on the head, neck and upper back. Force the chin in and the head and neck back. This exercise should be done explosively.

Do not confuse this exercise with the close grip upright row, which works some different muscle groups.

Exercise 6: THE DEADLIFT

Method

Stand with feet shoulder width apart, toes pointed slightly outward, with feet under the barbell, so that the shins are just behind the bar. Bend

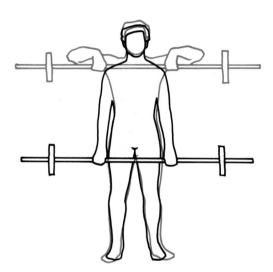

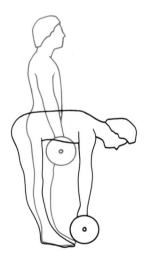

Figures 21 23. Upright Row (wide grip).

Figures 21-24. The Deadlift.

Figures 21-25. The straight legged dead lift.

Exercise 5: UPRIGHT ROW (WIDE GRIP)

Method

Take a shoulder width grip on the barbell and stand upright with the feet shoulder width apart, the bar hanging at thigh level, arms fully extended, knuckles to the front. Pull the bar straight up as high as possible, with the elbows kept higher than the hands. Lower the bar to the starting position.

at the knees and hips and grip the bar with the hands shoulder width apart (overgrip). Straighten the knees, hips and back to bring the body to an upright position, and the bar to the mid thighs. Lower the bar to the ground in a controlled movement.

Important Points

Concentrate throughout the lift, but especially

at the start and in the final phase of the lift, on ramming the chin into the neck and on pulling back with the head, neck and back. Do not worry about the position of your back when preparing to start the exercise.

It is possible to handle heavy poundages in this exercise. However, you should begin very cautiously to accustom your back to the movement.

After you have become proficient at this lift, and the muscles at the back of the thigh (the hamstrings) and the calves have had some stretching and strengthening, you can occassionally vary the exercise by, performing it with the knees straight throughout, bending entirely from the hips. This exercise is the Straight Legged Dead Lift. Care must be taken not to use too heavy a weight in this movement.

Exercise 7: POWER SNATCH

Method

With feet about hip width apart, bend at the hips and knees and grip the bar with the hands just over shoulder width apart. Pull the bar in one explosive continuous movement, to arms length overhead. Lower the bar to the ground in a controlled manner, or to mid-shin height if the weight is very light and the disc size is small.

Important Points

Keep the bar close to the body throughout the lift. The upwards movement should be explosive, the downwards movement controlled. Pause briefly between repetitions.

Exercise 8: BENT OVER ROW

Method

Bend over at the hips so that the back is flat and horizontal, with the knees slightly bent. The feet should be shoulder width apart. The bar should be held with an overgrip, the hands wider than shoulder width apart. Pull the bar up from arms length until it touches the chest, making sure that you maintain the bent over position. Lower the bar to arms length.

Important Points

The exercise should be done in a controlled manner, taking approximately one second to raise, and one second to lower the bar. Keep the chin tucked in, back flat and knees bent throughout the exercise.

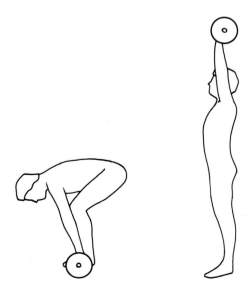

Figures 21-26. Power Snatch.

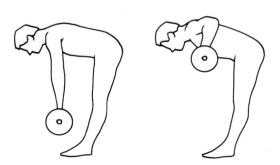

Figures 21-27. Bent over row.

Figures 21-28. Good Morning Exercise.   *Figures 21-29. Squats.*

Exercise 9: GOOD MORNING EXERCISE

Method

Place a light bar behind your shoulders, resting across your back. Stand with the feet shoulder width apart and the knees slightly bent. Bend forwards from the hips with the back flat until the back is almost horizontal, return to the upright position.

Important Points

This exercise must be done in a controlled manner, taking approximately $1^1/_2$ seconds to rise. The exercise puts considerable strain on the back, buttocks and hamstrings, so commence with an empty bar or very lightly loaded bar and progress cautiously. Do not let the bar roll up on to your neck during the performance of the exercise.

Exercise 10: SQUATS

If you are doing this exercise at home without the use of a squat stand, or the help of two assistants, then only a weight which you can comfortably lift over your head on to your shoulders can be used.

Method

Stand with the bar supported across the back, slightly below shoulder level, feet approximately hip width apart, toes turned slightly out. Keeping your back as flat as possible but inclined forward, bend the knees until the top of the thighs are parallel to the ground. Return to the upright position.

Important Points

This exercise must be performed in a controlled manner, taking about 2 seconds coming up. Under no circumstances should you go down so fast and so far that you bounce at the bottom of the movement, as this may injure the knee joint. However, as you become accustomed to the exercise a deeper squat can be performed, provided you do not bounce at the bottom. The deeper position when reached in a controlled manner, exercises the muscle and joints through a wider and more beneficial range than the parallel position.

It will be necessary for some people to perform this exercise with their heels raised between 2-5cms on a wooden block. Only those with extreme ankle and hip flexibility will be able to perform this exercise correctly, without using a block. It will be possible for most people to make rapid progress in this exercise and lift fairly heavy weights, when a squat stand will become essential.

EXERCISES FOR SIDEWAYS CURVE OF THE SPINE (SCOLIOSIS)

(a) *Scoliosis due to true shortening of one leg,* which may be the result of previous injury to the bone (fracture) or infection of the bone (osteomyelitis) or inherited. Raise the whole length of the shoe the appropriate amount to balance leg lengths, then do routine posture exercises.

(b) *Scoliosis due to apparent shortening of one leg* (pelvic tilt)

As has been explained in Chapter 12, apparent shortening of one leg occurs when the muscles in one leg which pull the leg towards the midline (adductors) are stronger than those which pull it away from the midline (abductors).

When apparent shortening occurs:

(1) The hip bone is higher on the apparently shorter leg. Check that your trousers or skirt rest evenly on your hip bones. Look in the mirror. If this shows one hip to be higher, that is the side of the apparently shorter leg.
(2) The outer edge of the shoe will be worn down on the apparently shorter side.
(3) There will be a tendency to stand with the knee locked back on the apparently shorter side and bent

on the other side.
(4) The hip joint on the apparently shorter side is more prominent. On either side of the body, midway between the back of the buttocks and the front of the pelvis, is a hollow (like a large dimple) in which can be felt the bone going into the hip joint.

On the apparently shorter side, this bone tends to stick out. On the other side the hollow is greater.

Exercise 11: (PELVIC TILT CORRECTION) Lateral (sideways) leg raise

Exercise only the apparently shorter leg, to correct apparent shortening.

(11A) AS A HOME EXERCISE WITHOUT RESISTANCE

Method

Stand upright, feet together holding the back of a chair placed with its side against a wall. Your shoulder will be against the wall. Raise the leg sideways as high as possible, keeping the trunk upright and the toes pointing forwards. Lower the

Figure 21-31. Outer edge of shoe worn down on apparently shorter side.

Figure 21-30. Check your skirt or trousers. The hip will be higher on the apparently shorter side.

Figure 21-32. Knee bent on apparently longer side.

Figure 21-33. Hip joint more prominent on apparently shorter side.

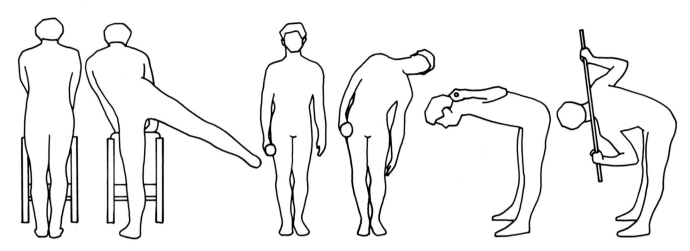

Figures 21-34. Lateral (sideways) Leg Raise.

Figures 21-35. Lateral (sideways) Trunk Bend.

Figures 21-36. Bent over trunk rotations.

leg. Perform the exercise in a slow controlled manner. You will not be able to raise your leg very high, if you do this exercise correctly.

Important Points

The leg you raise must be the apparently shorter one. The foot you raise must point to the front throughout the exercise. If you turn the foot out sideways, you will be exercising the wrong muscles.

If you do not have a suitable high-backed chair, use an open door for support.

Schedule (11A)

For the very frail and elderly, spend three weeks on Exercises 1(a), 2(a) and 3(a) before attempting this exercise and then only try with supervision.

First Week:
Do the exercise 8 times. Rest for one minute. Repeat the exercise 6 times. Rest for one minute. Repeat the exercise 4 times.

Second Week:
Do 10 repetitions. Rest. 8 repetitions. Rest. 6 repetitions.

Third Week (first week for the fit):
12 repetitions. Rest. 10 repetitions. Rest. 8 repetitions. Further improvement is better obtained by using resistance rather than increases in the number of repetitions.

(11B) AS A HOME EXERCISE WITH RESISTANCE

Attach weight, as described in Postural Exercise 1(b) to the leg, just below the knee.

Method

Perform the exercise as in Exercise (11A), slowly and deliberately, to prevent the weight banging against your leg.

Schedule (11B)

The more powerful may start with five kilograms, the rest with one kilogram. Concentrate on doing the exercise perfectly before attempting to increase the weight.

Follow Schedule (11A), increasing the weight when the 12, 10, 8 repetition pattern is reasonably easy. The maximum weight that can be reached is limited by the difficulty in attaching the weight to the knee. However, the exercise can be made more strenuous by attaching the weight to the ankle.

(11C) AS A GYMNASIUM EXERCISE

Use a low pulley system attached to the knee or the ankle. Follow Schedule (11A) beginning with a light weight and increasing the weight when the 12, 10, 8 pattern of repetitions can be performed with ease.

> **When Exercise 11 is used as a treatment for osteoarthritis of the hip both legs should be exercised but emphasis placed on the worse hip.**

Exercise 12: LATERAL (SIDEWAYS) TRUNK BEND

A HOME OR GYMNASIUM EXERCISE WITH RESISTANCE

Method

Stand with feet shoulder width apart. Hold a weight in the hand of the same side as the apparently shorter leg. Bend the trunk sideways, away from the apparently shorter leg. Return to the starting position.

Important Points

Do not let the trunk sway forwards or backwards. Do not let the hip sway sideways, towards the arm with the weight. Perform the exercise slowly with control.

Schedule

The weight used may be any convenient household equipment e.g. a pound jar of jam, a bottle containing fluid, or a bucket with the ability to increase the amount of fluid. An adjustable dumbell is the best equipment.

The very frail should start with one pound (0.5kg) the majority with 5 Kg and the strong with much more.

First Week:
8 repetitions. Rest. 6 repetitions. Rest. 4 repetitions.

Second Week:
10 — 8 — 6 repetition pattern

Third Week:
12 — 10 — 8 pattern.

Fourth Week:
Increase the weight, using the 12 — 10 — 8 repetition pattern.

EXERCISE FOR ROTATED SPINE

For those whose work requires repeated rotation of the spine in one direction only, e.g. lifting heavy bales of wool, the worker should endeavour to spend equal time working to either side. If this is not possible, Exercise 13 will help regain and maintain muscular balance. For sportsmen with a similar problem, e.g. discus and shot throwers, cricket fast bowlers, baseball pitchers, Exercise 13 will not only prevent later postural and arthritic problems, but will increase flexibility and performance.

Exercise 13: BENT OVER TRUNK ROTATIONS

Method

Load a bar so that there is weight on one end only. Stand with the feet shoulder width apart, with the bar resting across the shoulders behind the neck. The weight should be on the opposite side to your turn, i.e. if you turn to the left in your work or sport, the weight should be on the right side of the bar. Bend forward at the hips until the trunk is horizontal (parallel to the ground) and try to keep the back straight.

Rotate the trunk towards the weighted side of the bar as far as possible, and then rotate back to starting position.

Important Points

The weight must be very secure, otherwise the weight will fly off. Secure the weight with two collars and double check them. Make sure that you will not hit anybody when you rotate the spine. Rotate the spine in one direction only. Do the exercise in a controlled manner, trying to rotate further each time. Do not rebound back from one repetition into the next repetition.

Schedule (13)

This exercise is only necessary and advisable for fit, strong and pain free individuals and should be performed approximately three times per week.
Load the bar with 5kg initially. Do 12 repetitions. Rest. 10 repetitions. Rest. 8 repetitions.
Increase the weight when performing these repetitions comfortably.

N.B. A variation of this exercise can be done with a pulley machine.

EXERCISES FOR OSTEOARTHRITIS IN NON-WEIGHT BEARING JOINTS

A. ELBOWS

Exercise 14: **ELBOW STRAIGHTENING (triceps extensions)**

Method

Take a weight in the hand. This may be either a jar of jam, a bucket containing some water, or a dumbell. Start with a weight which can be managed with comparative ease, holding the weight with the knuckle pointing backwards.
Bend forward at the hips and rest the other arm on a table or bench. Keep the upper arm against the side of the body, but allow the forearm to hang vertically down. Raise the weight backwards by straightening the elbow. Then lower the weight again.

Important Points

Initially, full straightening may be prevented by pain or stiffness. However, continue with the exercise and eventually full straightening should occur.

Schedule (14)

Use the same schedule as in Exercises 4-10

B. SHOULDER JOINTS

Perform Exercise 4 and, in addition,

Exercise 15: LATERAL (SIDEWAY) ARMS RAISE

(This exercise will also correct a "frozen shoulder" resulting from calcium deposited in injured muscles and joint capsule of the shoulder joint.)

Method

Stand with feet shoulder width apart and a weight in each hand touching the outside of the thigh. Keeping the elbows straight and palms turned inwards, raise the arms sideways above the head. Lower the arms in the reverse direction.

Important Points

It is vital that this exercise is performed with control, both going up and coming down, taking up to 2 seconds in each direction.
Initially, raise the arms until restricted by pain or immobility. The exercise will increase the mobility and reduce the pain, until a full range of movement is achieved. Exercise both arms together, even if only one shoulder is affected. This will prevent sideways twisting of the trunk.

Schedule (15)

Start with an 8 — 6 — 4 repetition pattern and increase to a 12 — 10 — 8 pattern. Maintain this pattern and increase the weights when the exercise can be performed with ease.

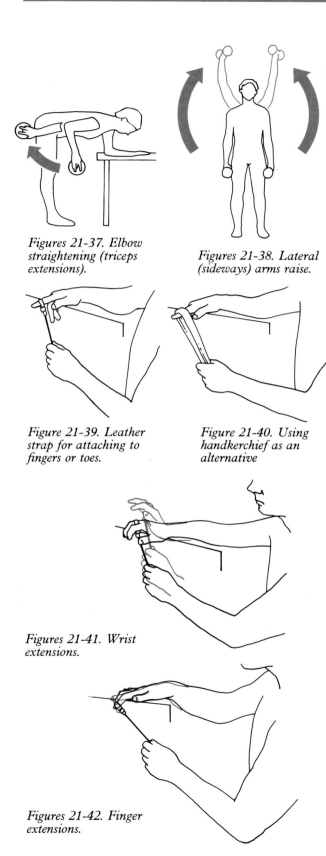

Figures 21-37. Elbow straightening (triceps extensions).

Figures 21-38. Lateral (sideways) arms raise.

Figure 21-39. Leather strap for attaching to fingers or toes.

Figure 21-40. Using handkerchief as an alternative

Figures 21-41. Wrist extensions.

Figures 21-42. Finger extensions.

C. WRISTS, FINGERS AND TOES

To exercise these joints I use a leather strap, which will fit over the fingers or over the toes. Fixed to this strap is a shoe lace. Pressure can be exerted on this shoe lace either by pulling with the opposite hand or by attaching a small weight. An alternative method is to roll up a handkerchief so that a strap is made.

Exercise 16: WRIST EXTENSION

Method

Fix the strap or handkerchief over the middle finger as close to the palm as possible. Pull down on the strap or handkerchief with the opposite hand. Bend the wrist backwards as hard and as far as possible, gradually easing off the pull of the opposite hand.

Schedule (16)

Do 10 repetitions at least twice a day, but work on this exercise at odd moments, as often as you like.

Exercise 17: FINGER EXTENSION

Method

To exercise a joint in any finger, place the strap beyond (i.e. the fingernail side) the joint, and try to straighten the joint against a pull in the opposite direction. Ease off on the opposing pull so that the finger can be straightened as much as possible.

Important Points

This exercise is also a successful way of treating Duyputren's Contracture, where the ring and little fingers become curled and will not straighten.

For 15 years, a friend who is a concert pianist, has used this exercise to regain the flexibility of his finger joints after he has been gardening, mixing concrete or doing other heavy labour.

Schedule (17)

Do 10 repetitions at least twice a day, but work at this exercise at odd moments, as often as you like.

Exercise 18: FINGER FLEXION

Typists frequently develop rigid joints at the end of their fingers. They need to persevere with exercises 17 and 18, to regain pain free, fully mobile normal sized joints.

Method

If the fingers will not bend fully, reverse the strap so that the pull is in the opposite direction. Now try to bend the appropriate joint as much as possible, while easing off on the opposing pull.

Schedule (18)

Do 10 repetitions at least twice a day, but work at this exercise at odd moments, as often as you like.

Exercise 19: TOE EXTENSION

Similar to finger extension

Exercise 20: TOE FLEXION

Similar to finger flexion

Exercise 21: ANKLE DORSI FLEXION

Flexibility of the ankle joint is important for posture, but vital for sport. By increasing the range of pulling the foot up, it will increase the range and power that the foot can be pulled down, to give drive. By stretching the calf muscles and the achilles tendons to their maximum, they are less likely to be injured, and shin splints will be avoided. This exercise is also vital for the rare condition of osteoarthritis of the ankle joint, when it should be practised for the first three weeks without using weights.

Method

Fix a weight, using the strap shown in Figures 21-8 and 21-9, to the underside of the foot near the base of the toes. Raise the foot by bending at the ankle. Lower the foot.

Schedule (21)

Commence with a 1kg unless very strong and flexible, when up to 5kg may be used. First week, 8 − 6 − 4 repetition pattern. Second week, 10 − 8 − 6 pattern. Third week 12 − 10 − 8 repetition pattern. Increase the weight when these repetitions can be done comfortably.

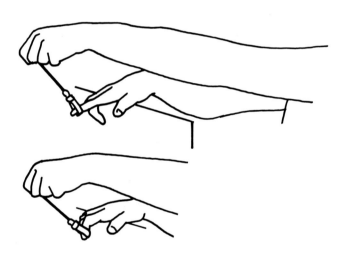

Figures 21-43. Finger flexions.

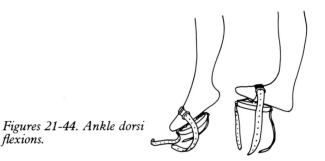

Figures 21-44. Ankle dorsi flexions.

Exercise 22: LEG RAISE

> **Whether you are lordotic or kyphotic, this exercise is vital during pregnancy. Commence the exercise as soon as you know that you are pregnant, and continue doing it until you go into labour.**

Method

Lie flat on your back, keeping the knee straight. Raise the leg to the vertical, then lower it, almost to the ground, repeating for the required number of repetitions.

Important Points

Perform the exercise at a steady controlled pace, both going up and down. If you are very weak, touch your heel on the ground between repetitions.

Schedule
(Repeat twice daily)

Some women will be stong enough to commence the exercise with both legs. If in doubt, use one leg only.

First Week:
Do 10 repetitions with the right leg. Rest for one minute. Do 8 repetitions. Rest. Do 6 repetitions. Repeat the same sequence with the left leg.

Second Week:
12 — 10 — 8 repetition sequences with each leg.

Third Week:
(First week for those who can commence with both legs)
6 repetitions with both legs. rest one minute. 4 repetitions. Rest. 2 repetitions.

Fourth Week:
8 — 6 — 4 repetition pattern.

Fifth Week:
10 — 8 — 6 repetition pattern

Sixth Week:
12 — 10 — 8 repetition pattern. Continue at this level until you go into labour, and then for a further month after delivery.

Exercise 23: PARTIAL SIT UPS

Lie on your back with your feet on the floor and your hips and knees bent. Press the small of your back down towards the floor and try to flatten your spine. Rest your fingers on the top of your thighs. This is the starting position. First bend your head forward, followed by your chest, with your fingers moving down your thighs to your knees. Uncurl back in the opposite direction until your head is back on the floor.

Important Points

Try to slowly uncurl your spine from the floor.

Figures 21-45. Leg Raise.

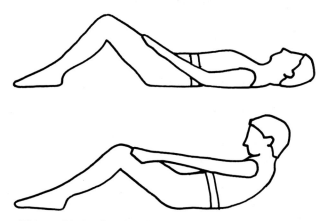

Figures 21-46. Partial sit ups.

Schedule
   (Twice daily)

First Week:
   8 repetitions. Rest for one minute.
   6 repetitions. Rest for one minute. 4 repetitions.

Second Week:
   10 — 8 — 6 repetition pattern.

Third Week:
12 — 10 — 8 repetition pattern.

EPILOGUE

Though you drive nature out with a pitchfork,
She will ever return.

Horace (65B.C. — 8 B.C.)

This book has been about posture. It has also been about evolution; about man's interference with evolution; and about the cost of this interference.

For millions of years in, and millions of years out, evolution has ground its way steadily along. It worries not what you call it — evolution, nature or God.

Its sole interest is whether it works. It simply obeys the laws of physics. If it does not work, if it does not improve the efficiency of life, sooner or later it is rejected.

The principles of evolution should be the prime concern in all our deliberations.

"All nature is but art, unknown to thee;
All chance, direction which thou canst not see;
All discord, harmony not understood;
All partial evil, universal good:
And, spite of pride, in erring reason's spite,
One truth is clear, Whatever is, is right."

Alexander Pope (1688-1744)

ELECTRIC MAN

This chapter is not related to posture, and you will find no practical benefit to your posture in it. However, it stemmed from my research into posture, out of which arose hypotheses that may prove to be very useful and may fascinate you as they have fascinated me.

THE ELECTRO-MAGNETIC HYPOTHESIS OF MUSCLE CONTRACTION

Believing that posture was the result of the relative tone of muscles acting upon joints, I was determined to find out exactly what muscle tone was, and how it was produced. During my research I developed a theory to explain muscle contraction that differs from the generally accepted theory.

A muscle consists of a muscle belly, with a tendon at either end attached to a bony surface. The tendons simply transmit the movement or force which the muscle generates. The muscle generates this force by shortening, or attempting to shorten. It appears to have no power to lengthen itself, nor to revert to its original length after contraction. It would appear that it is pulled back to its previous length and shape by the pull of the opposing muscle, when that contracts.

If the muscle belly is dissected down to its smallest part we come to the contractile unit, called a sarcomere, where the contracting process takes place. Much of our knowledge of the sarcomere comes from the brilliant work of H.E. Huxley at Cambridge. Millions of these minute cylindrical, contractile units are attached to each other lengthwise, and also side by side. The more units we activate along the length of the muscle, the shorter it will become. The more units we activate across the thickness of the muscle, the more powerful will be the contraction.

This simple mechanism can lift a feather or a hundred pounds with smoothness, ease and control.

The problem is, how does the sarcomere (the contractile unit) contract? Huxley proposed a Sliding Filament Theory.

> **I believe that it acts like a double solenoid — a simple electric machine.**

A solenoid is a closely packed coil of wire through which a current of electricity is passed.

As a current 'I' is passed through the coil of wire, the coil acts as a bar magnet with poles at either end, so that a metal bar will be attracted into the centre of the coil.

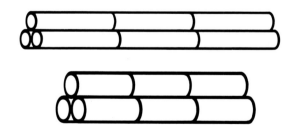

Figure A-1. Illustration showing nine of the millions of contractile units which make up a muscle — before contraction.

Figure A-2. Nine contractile units — after contraction.

The solenoid has many uses. If we fix our Bar B, to plate T, and arrange a solenoid coil either end of the bar and allow the coils themselves to move (Figure A-6) when we pass a current simultaneously through both coils, these coils will snap together (Figure A-7).

I maintain that this double solenoid action produces the shortening, i.e. the contraction in the muscle unit (sarcomere).

The electron microscope has shown this central bar which is known as a myosin filament. It has also demonstrated plates at either end of the sarcomere and that, when an end plate is stimulated electrically it moves towards the middle of the central bar. The electron microscope also shows some slender rods (actin filaments) fixed to the end plate, and cross connections (cross bridges) between these slender rods and the central bar.

The generally accepted theory is that these cross bridges, broken and distorted by the technique of preparation of the muscle material for the electron microscope, are slender connections linking the thick myosin filament to the thin acton filament, and that they produce the contraction in the sarcomere.

I believe that these cross-connections are the remnants of solenoid coils, and that the thin rods either assist the mechanism or are simply a thickening where adjacent coils cross each other, and the unexplained connection remnants between adjacent actin filaments are also part of the solenoid coils. There are not only cross connections between adjacent myosin and actin filaments, but there are also cross connections between adjacent actin filaments, and no satisfactory explanation is given for their presence. However, if the double solenoid theory is correct these cross connections would be part of the solenoid coils.

It is also known that when the end plates are stimulated electrically, and the unit contracts, a chemical reaction takes place which releases electrons. Whatever the method of contraction, evolution has had many problems to solve concerning the contraction of the sarcomere.

(1) There is a rapid build up of heat on contraction which has to be removed rapidly.

(2) The contents of the cylindrical contractile unit is mainly fluid. When it contracts this fluid cannot be compressed, so it has to escape somehow.

(3) Material for the production of energy must be brought into the unit quickly, and waste products

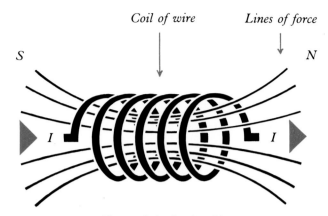

Coil of wire   Lines of force

S   N

I   I

Figure A-3. A solenoid.

No current

Figure A-4. A solenoid, and a metal bar — No current.

I   I

Figure A-5. Metal bar attracted, when current is passed.

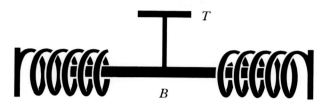

Figure A-6. A fixed metal bar, with a solenoid at either end of it.

Figure A-7. When current is passed through the coils they snap together.

End plate   Cross connections

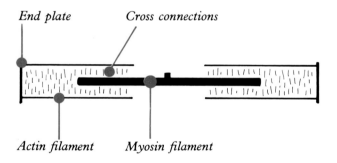

Actin filament   Myosin filament

Figure A-8. The sarcomere contractile unit showing cross bridges.

Cross connections — always present but never satisfactorily explained

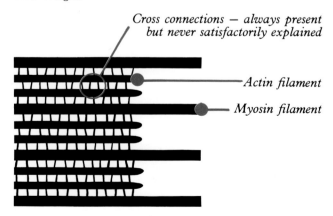

Actin filament
Myosin filament

Figure A-9. Part of three adjacent sarcomeres, to demonstrate cross connections.

removed as quickly.

(4) The outer coat must shorten and then regain its length — or change its shape.

I consider that all these problems have been solved beautifully and in a simple way with an hydraulic pump.

There are sacks (A) along the outer side of the length of the sarcomere, connected to the centre of the sarcomere. And other sacks (B) at the ends of each sarcomere, connected to two consecutive sarcomeres (Figure A-10). When the sarcomere contracts (Figure A-11) some of the fluid is forced out of it into the sacks, carrying heat and waste products to the blood vessels (capilliaries), and forcing the outer coat sideways. When the sarcomere is pulled back to its full length, the fluid in the sacks is sucked back into the sarcomere, bringing with it oxygen and nutrients from the blood vessel, and the side walls can lengthen.

With modern technology, it seems feasible that the contraction of human (and animal) muscle can be duplicated. Such an imitation muscle could be adapted for those who have lost a limb, or who have muscle paralysis.

For industry, an artificial muscle has innumerable uses. Most machinery is worked by a pushing movement. There is little which pulls. In the burgeoning robot industry the use of artificial muscle would seem ideal, as it can be designed for extreme delicacy or immense power; it is silent, does not pollute, has great range, mobility, flexibility and control, it is simple to computerise, and requires no power unit other than connection to an electrical supply. If the double solenoid principle is used in reverse, tidal or wave power could be used to generate electricity. Already superconductors are being manufactured from organic materials, and it is probable that in the voluntary muscle of the animal kingdom will be found the best of superconductors. In time, the secrets that the animal and vegetable world use to produce energy may be revealed, enabling us to convert glycogen and other foodstuffs into electrical energy.

THE ELECTRO-MAGNETIC HYPOTHESIS OF CELL DIVISION AND CANCER PRODUCTION

The double solenoid hypotheses only works in the contraction of voluntary muscles, i.e. the muscles we can contract of our own conscious volition. Many muscles e.g. those in the walls of our blood vessels, or the intestines, are controlled subconsciously, and these muscles do not have the central bar (myosin filament), so cannot contract in this way.

Brilliant work by Bernard J. Panner and Carl R. Honig at the University of Rochester, New York, has led me to believe that involuntary muscle is composed of continuous, organic electrical coils. When they are electrically stimulated they contract but produce less dynamic contraction than voluntary muscle. These coils can be curved to produce the circular contraction necessary in blood vessel or intestine walls, whereas the rigid central bar (myosin filament) of the double solenoid prevents curved or central contraction.

Heart muscle is unique in that it requires the powerful contraction that the myosin filament provides, but it must contract concentrically. It has partially solved this problem by arranging some of the muscle in a spiral. But I suggest that much of the heart muscle is built like a jigsaw of interlocking pyramids, with their bases on the outside and their top portions removed to form the hollow chamber of the heart.

However, there is such an enormous number of pyramids that the outside layer does not appear angled, but appears as a gradual curve. The sarcomere units are equal in size. As there must be more of these units in a row close to the base of a pyramid than a row nearer the centre, then, instead of the units lying exactly side by side, as we see in voluntary muscle, they must have a stepped ladder effect, in heart muscle. Even in the spiral part of heart muscle this step-ladder effect must occur. And this is what we find on microscopic examination.

After a contraction the sarcomeres in the heart muscle would be returned to their uncontracted state by the stretching effect of the thick, viscous blood returning to the chambers of the heart.

Besides their use in involuntary muscle, I suggest that nature uses organic electrical coils for a whole variety of movement. The wave motion of cilia (microscopic hair like structures on the lining of the sinuses and lungs) and flagella (tail like structures that propel some organisms in water) and the sudden straight-line movements of particles inside cells are produced by similar structures known as microtubules. I contend that these

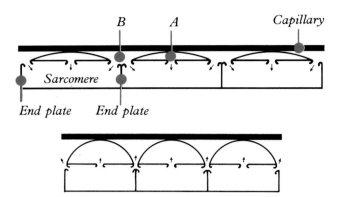

Figure A-10. Illustration of three sarcomeres to demonstrate fluid movement into these sarcomeres, as they are stretched, and, Figure A-11, as they contract.

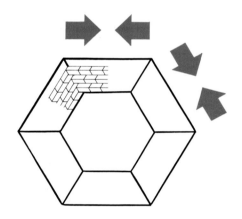

Figure A-12. Scheme of relaxed heart muscle.

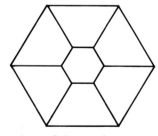

Figure A-13. Scheme of contracted heart muscle.

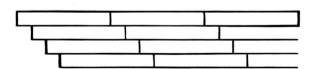

Figure A-14. Step ladder effect of sarcomeres in heart muscle.

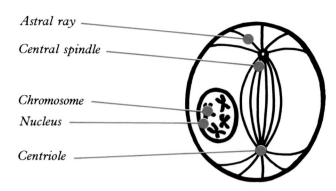

Astral ray

Central spindle

Chromosome

Nucleus

Centriole

Figure A-15. Start of division of a cell, with the chromosomes in the nucleus splitting into two.

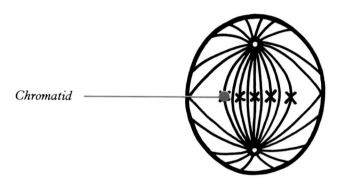

Chromatid

Figure A-16. Chromosomes arranged on central spindles.

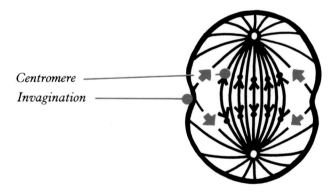

Centromere

Invagination

Figure A-17. The cell being pulled apart by astral rays.

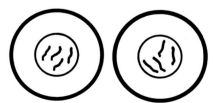

Figure A-18. Division of cell completed to form two new daughter cells.

microtubules are organic electrical coils. Microtubules are essential for the division of cells — the most vital activity of life. It is the uneven and uncontrolled division of cells that is termed cancer.

When a cell prepares to divide, star like bodies (the astral rays) appear in each half of the cell and begin to move apart. These astral rays are connected to each other by the central spindle. Both the central spindle and the astral rays are made of microtubules. The chromosomes inside the nucleus divide into two halves (chromatids) and then the nucleus disintegrates.

The chromosomes then arrange themselves on the central spindles, midway between the astral rays.

> **I contend that the elements that form the astral rays and central spindle are actually organic electrical coils, which are responsible for cell division.**

The central spindle coils which have chromosomes attached, pull the chromosome halves towards their nearest astral ray. The chromatid half has a central plate (the centromere), and I believe that this enhances the electrical pull. The other central spindle coils hold the astral rays in position, and stop them separating further.

The astral rays which are connected towards the centre of the cell start to pull the cell in half to divide it in the middle.

In order for astral rays to pull the invagination inwards to split the cell in half, the astral ray centres move further towards the periphery of the cell. This movement could occur by reducing the pull of the centre spindles, or by increasing the pull of the astral rays which are in line with the two astral ray centres. When the cell has split in two, the chromotids become whole chromosomes and they form part of the nucleus of the new daughter cell.

If my hypothesis is correct, a series of organic electrical coils (central spindle) pull the divided chromosome into each half of the cell, and similar coils (astral rays) pull the wall of the cell inwards across its middle splitting the cell in two. After division these coils either disintegrate chemically, or remain to provide a movement force inside the cell. It is probable that as coils disappear in some cells, they form in others so that a constant electromagnetic balance can be maintained in the organism.

If we wish to control the wild cell division of cancer, treatment research based on my theory of electromagnetic cell division should concentrate on the normalisation of the minute electrical variations that exist on the surface of and inside the cells, by first understanding the nature of the bodies pattern of interlocking electromagnetic fields, and the causes underlying their variations. Given time and sufficiently refined measurement techniques, we should be able to determine exactly what is normal and how electromagnetic forces are affected by changes in the organism and how they can be modified for repair.

One possible line of research would be to examine the electromagnetic variations produced on the surface of, and inside a cell after it has been penetrated by a virus; and another would be to examine, in the same way, the consequences of radiation at various levels.

RECHARGING THE BRAIN

In controlling our body, the brain sends a series of static electrical charges along the nerves to the various parts. There must come a time when the available store of static electricity in the brain is depleted. I suggest it takes about four minutes to use up this store — the approximate time it takes for the death of the brain, after the circulation to it has stopped.

WHAT RECHARGES THE BRAIN?

There is one simple and very plausible explanation. As the red blood cells in the blood brush against the walls of the blood vessels, particularly in the minute blood vessels (the capillaries), they pick up a static electrical charge in the same way that combing the hair will produce a charge. When these red cells reach the capillaries in the brain, their charge is then transferred to the nerve cells by special brain cells called astrocytes. These extremely numerous cells not only have connections to the nerve cells (neurons) in the brain, but also have numerous "end-feet" (or "vascular feet") which are fixed on to the outside wall of the capil-

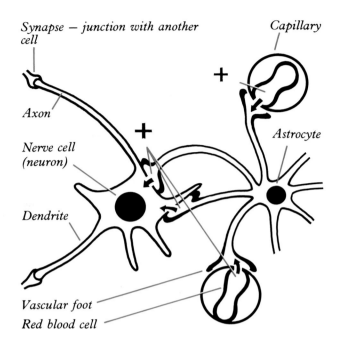

Figure A-19. Diagram to illustrate transfer of electrons from red blood cells in the circulation to nerve cells in the brain.

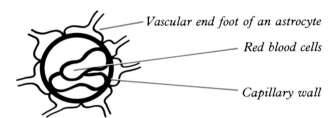

Figure A-20. Diagram showing capillary in brain completely surrounded by end feet of astrocytes, to pick up electricity to recharge the brain.

laries in the brain.

In fact, the capillary is almost completely covered by these "vascular feet".

The sensitivity of astrocytes to potassium ion concentration, the high concentration of potassium ion in red blood cells, and the part the astrocytes play in the "blood-brain barrier", all indicate that the astrocytes' main purpose is to "pick up" the electrical charges from the red blood cells and transfer them to the nerve cells of the brain.

The realisation that the brain has to be recharged, and that the charge comes from blood movement (i.e. circulation) has tremendous significance to our well-being, especially to ageing — and to specific and general disease processes.

ELECTRIC MAN

The brain discharges an electrical signal along the nerves to muscles. This stimulus causes the muscle to contract, using the electro-magnetic principle of the double solenoid. Thus the muscle performs work for electrical man. Food ingested by man is needed to provide the energy for the next contraction of the double solenoid. The contraction of the muscle forces the blood along the capillaries. The brain also sends an electrical stimulus to the heart to contract, and this contraction is again performed by electrical forces of the double solenoid. The pumping of the heart ensures the movement of the blood along the capillaries. The contact of the red blood cells on the capillary walls produces static electricity, which is released to the brain (to recharge it) by the astrocyte cells.

Thus Man, in fact all animal life (and possibly plant life) is built around an electrical circuit. The secret of a living brain, i.e. of life itself, is an adequate supply of blood to recharge the brain. This supply comes from muscular contraction, from physical activity, from movement.

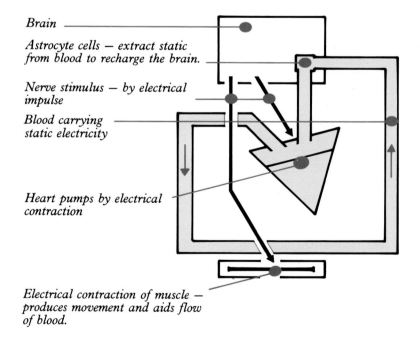

Brain

Astrocyte cells — extract static from blood to recharge the brain.

Nerve stimulus — by electrical impulse

Blood carrying static electricity

Heart pumps by electrical contraction

Electrical contraction of muscle — produces movement and aids flow of blood.

Stillness is stagnation and death. Movement is life.